Springer Tracts in Mechanical Engineering

Springer Tracts in Mechanical Engineering (STME) publishes the latest developments in Mechanical Engineering - quickly, informally and with high quality. The intent is to cover all the main branches of mechanical engineering, both theoretical and applied, including:

- Engineering Design
- Machinery and Machine Elements
- Mechanical structures and Stress Analysis
- Automotive Engineering
- Engine Technology
- Aerospace Technology and Astronautics
- Nanotechnology and Microengineering
- Control, Robotics, Mechatronics
- MEMS
- Theoretical and Applied Mechanics
- Dynamical Systems, Control
- Fluids mechanics
- Engineering Thermodynamics, Heat and Mass Transfer
- Manufacturing
- Precision engineering, Instrumentation, Measurement
- Materials Engineering
- Tribology and surface technology

Within the scope of the series are monographs, professional books or graduate textbooks, edited volumes as well as outstanding PhD theses and books purposely devoted to support education in mechanical engineering at graduate and post-graduate levels.

Indexed by SCOPUS and Springerlink. The books of the series are submitted for indexing to Web of Science.

To submit a proposal or request further information, please contact: Dr. Leontina Di Cecco Leontina.dicecco@springer.com or Li Shen Li.shen@springer.com.

Please check our Lecture Notes in Mechanical Engineering at http://www.springer.com/series/11236 if you are interested in conference proceedings. To submit a proposal, please contact Leontina.dicecco@springer.com and Li.shen@springer.com.

More information about this series at http://www.springer.com/series/11693

Jorge Angeles · Damiano Pasini

Fundamentals of Geometry Construction

The Math Behind the CAD

Springer

Jorge Angeles
Department of Mechanical Engineering
McGill University
Montreal, QC, Canada

Damiano Pasini
Department of Mechanical Engineering
McGill University
Montreal, QC, Canada

ISSN 2195-9862 ISSN 2195-9870 (electronic)
Springer Tracts in Mechanical Engineering
ISBN 978-3-030-43133-4 ISBN 978-3-030-43131-0 (eBook)
https://doi.org/10.1007/978-3-030-43131-0

This Springer imprint is published by the registered company Springer Nature Switzerland AG
The registered company address is: Gewerbestrasse 11, 6330 Cham, Switzerland

Beware of designers who walk around without paper and pencil.

Ancient Chinese proverb.

Preface

The purpose of the book, in a nutshell, is to provide both the beginner and the experienced CAD user with *the math behind the CAD*. It is expected that, with the geometry tools introduced here, the reader will be able to exploit CAD software to its fullest extent. In fact, the book should allow the reader to go *beyond what CAD software packages offer in their menus*.

Chapter 1 includes a summary of facts from basic Linear and Vector Algebra, most of which the reader may have covered either in high school or in undergraduate math courses. The chapter also includes some new items, such as the 2D form of the vector product. In fact, while the vector product is defined solely in 3D, it is needed in some 2D applications, as made apparent in this chapter. The 2D form is not common in mathematics textbooks, although it is a useful tool that helps to streamline many calculations, as the reader will find here. One more novelty, not covered in either high school or undergraduate math courses, is the manipulation of block matrices and vectors. This material is introduced to help the reader compute inverses and determinants of $n \times n$ matrices, for $n > 3$ using the corresponding expressions for 2×2 and 3×3 matrices. These calculations become handy in Chap. 4, devoted to affine transformations.

Chapter 2 includes the study of the relations among points, lines, and curves in the plane. Here, the difference between curves representing functions and their geometric counterparts is emphasized. Plotting the former is quite a different task from plotting the latter. Scientific software makes the difference, in that it uses different commands to plot one or the other. The former are representable by means of *explicit functions*, e.g., $y = f(x)$, the latter by *implicit functions*, e.g., $f(x, y) = 0$. Geometric concepts are studied in order of growing complexity, from points to lines, then algebraic curves and, finally, non-algebraic and free-form curves. The simplest algebraic curves, conics, are given due attention. However, these are too limited for practical engineering design applications. For example, they cannot produce a smooth fillet to join the two lines of a corner with tangent and curvature continuity, which is essential to avoid stress concentrations in mechanical design. For this reason, Lamé curves are introduced in this chapter.

Chapter 3 is devoted to geometric objects in 3D, namely, points, planes, lines, and surfaces. Of the latter, only quadrics are studied, to keep the discussion at an elementary level. It is shown that, with these simple surfaces and planes, more complex surfaces can be generated by means of *Boolean operations*. These are at the basis of CAD software. Furthermore, as illustrated in Chap. 4, highly complex surfaces, like those of screws, can be readily generated by means of affine transformations.

The concept of *affine transformations* is indeed a key component of Chap. 4, which includes applications of these transformations to the synthesis of curves and surfaces that could be extremely cumbersome to produce otherwise. Again, affine transformations lie at the core of CAD software. Understanding these transformations should help the reader best use the software tools available in the market.

In this book, catering to various disciplines—engineering, graphic design, animation, and architecture—an attempt is made to keep the material discipline-independent, while including some examples of interest to various disciplines. Furthermore, the book can be used to complement lectures on CAD, at the undergraduate level. In this light, the references included in footnotes are not mandatory reading. They are included for the benefit of both the lecturer and the curious, diligent student wanting to go deeply in certain subjects.

A crucial point that led to the production of the Lecture Notes, and hence to this book, was the input received from Axel Pavillet, Ph.D., alumnus of the French *Ecole Polytechnique*.[1,2] In the early 2000s, Dr. Pavillet was an Instructor of the course on *engineering drawing* offered by the Department of Mechanical Engineering at McGill University. The syllabus of this course included, as was usual in those days, a substantial component of *Descriptive Geometry* (DG). Dr. Pavillet's point was that DG was completely out of place since the end of the twentieth century, given that computer-graphics software was already available in the 80s. Lengthy exchanges with Dr. Pavillet and other instructors finally led us to the idea that the course in question needed an in-depth revision, with classical DG replaced with fundamental concepts of geometry and linear algebra. Thus came this book to fruition.

This book is the result of teamwork. We thank all the students, undergraduate and graduate, who contributed with their work to bring the book to its current form. We acknowledge especially the work of two Ph.D. students who were instrumental in both the first phases of the book editing and producing graphic material—photos and drawings—that led to the final version. The highly professional work done by Vikram Chopra and Wei Li is dutifully acknowledged, with our most sincere thanks. The editing work leading to the final version was conducted by Dr. Bruno Belzile, Postdoctoral Fellow at the *Robotic Mechanical Systems Laboratory* (RMSLab). The RMSLab is affiliated with both the *Centre for Intelligent Machines* and the *Department of Mechanical Engineering*, McGill University, in Montreal. Dr. Belzile provided also technical support in the graphics work that made possible

[1]Asancheyev (2002).

[2]Pavillet (2011)

the completion of this highly demanding project. Our most sincere acknowledgement to Dr. Belzile for his diligent and excellent work.

The support of both the *NSERC Design Engineering Chair* "Design for Extreme Environments" at McGill University, in the period 2003–2008, and the *McGill Teaching and Learning Improvement Fund* is dutifully acknowledged. In fact, NSERC's support through the aforementioned Chair made possible the initiation of the manuscript that evolved from a set of *Lecture Notes* to the document that eventually led to the present book. Last but not least, the support received from McGill University through a *James McGill Professorship*—with the motto "Towards a Theory of Engineering Design"—to the first author in the period 2003–2017 was instrumental in providing resources that played a significant role in the completion of this book. Besides the foregoing support received from McGill University, we also acknowledge the permission to reproduce the McGill University crest and its distorted versions in Chap. 4.

Any suggestions for improving or any reports of typos or inaccuracies are most welcome.

Montreal, Canada
December 2019

Jorge Angeles
angeles@cim.mcgill.ca

Damiano Pasini
damiano.pasini@mcgill.ca

References

Asancheyev B (2002) Épures de Géométrie Descriptive Concours d'entrée à l'École Normale Supérieure. Hermann Publishers, Paris, pp 11–13

Pavillet A (2011) Replacing de(ad)scriptive geometry..., https://ojs.library.queensu.ca/index.php/PCEEA/article/download/4059/3993

Contents

List of Figures

Chapter 1
Introduction to Geometry Construction

1.1 Book Overview

The design process starts with a need, as spelled out by the client. In engineering design, as well as in other design areas, the need is described by the client in rather ambiguous, fuzzy, sometimes contradictory terms. After a series of exchanges between client and designer, be this an engineer, an industrial designer, or an architect, the need is formulated in terms of a list of *functional requirements*, with some specific features that are spelled out as *design specifications*, or *specs* for brevity.

1.1.1 Free-Hand Sketches

Once the functional requirements and design specifications are agreed upon by client and designer, the latter produces *free-hand sketches* of some design alternatives. The importance of free-hand sketching skills in design cannot be overstated. The quality of the final design solution is highly dependent upon how the designer can communicate her or his ideas not only to the client and to other professionals, including her or his design team, but also to herself or himself in an unambiguous, concise, and clear way. Developing basic sketching skills, however, is not the subject of the book.

Shown in Fig. 1.1 is a free-hand sketch produced by a professional designer *to embody* the design of a mechanism housing, to serve as a means to protect the mechanism, displayed in Fig. 1.2 as mounted on a jig, and to support firmly its various moving parts. Not only this, thc housing is also intended to protect humans from accidentally putting their hands in the way of the moving parts of the mechanism (Fig. 1.3).

Every engineering designer pays attention to the *lettering* that accompanies invariably every engineering free-hand sketch, as illustrated in Fig. 1.1. Poor lettering may be illegible, and hence conveys no information; at worst, poor lettering may be

J. Angeles and D. Pasini, *Fundamentals of Geometry Construction*, Springer Tracts in Mechanical Engineering, https://doi.org/10.1007/978-3-030-43131-0_1

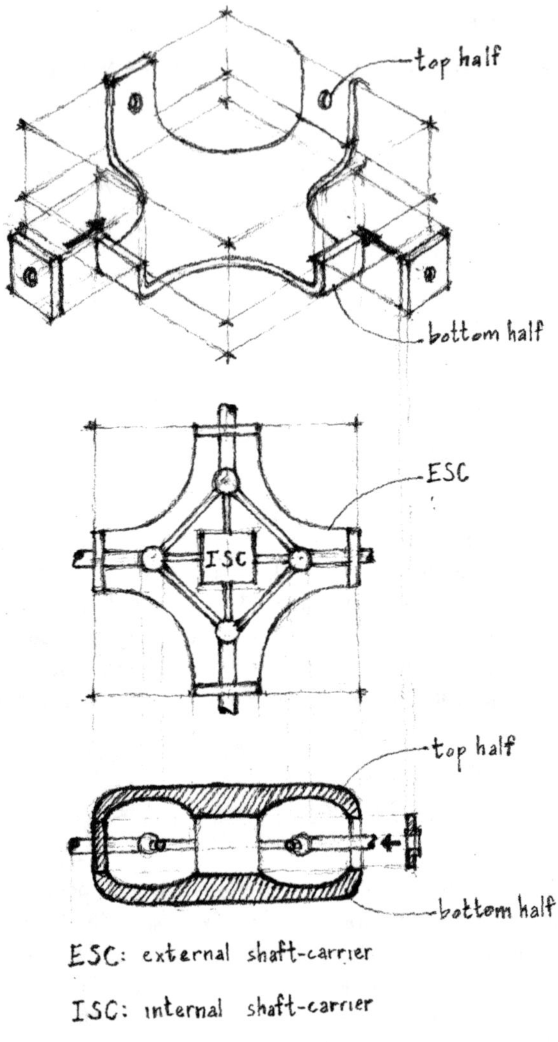

Fig. 1.1 The free-hand sketch of a solution alternative to the problem of designing a mechanism housing (Sketches produced by Juan Fernández González, B.Sc. Arch.)

confusing, and hence conveys the wrong information. Lettering affects the quality of a free-hand sketch.

The book aims to develop the analytical skills required in producing accurate, unambiguous engineering drawings, to be used by other design professionals and technicians for manufacturing or construction. The content relies on elementary knowledge of algebra and geometry as prerequisites. The main objective of the book is to teach the reader the "math behind the CAD," where CAD stands for *Computer-Aided Design*, a technology that frees the designer from the routine tasks of design drafting and modeling. The reader is introduced to CAD technology via the fundamentals.

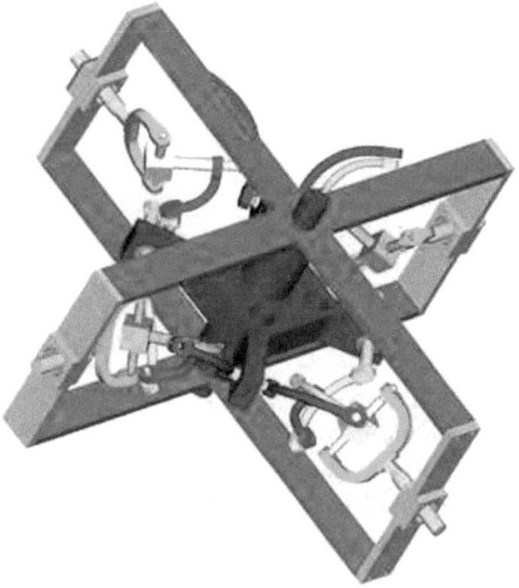

Fig. 1.2 The geometric modeling of a complex mechanism, mounted on a jig, in need of a housing

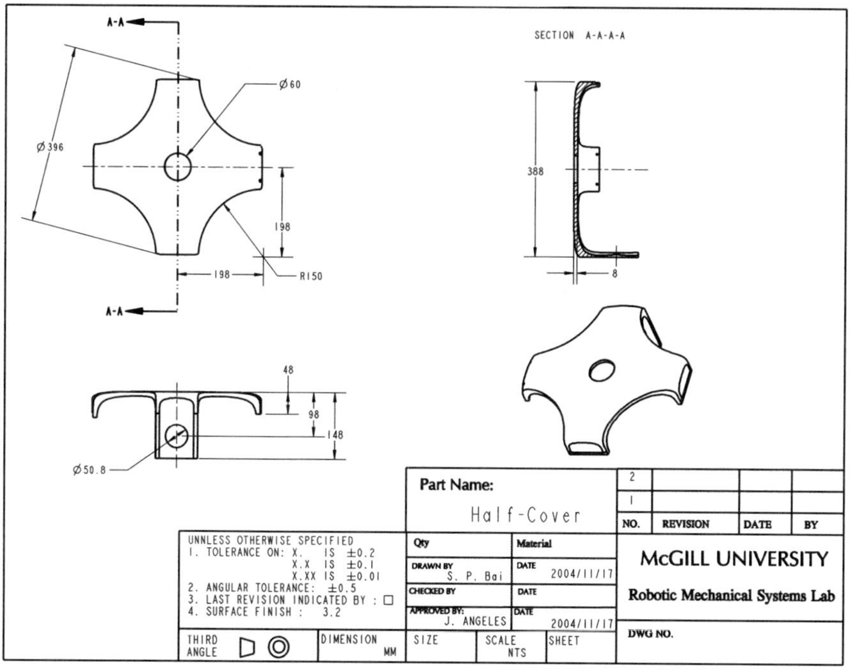

Fig. 1.3 A typical manufacturing drawing of a part of the housing sketched in Fig. 1.1, with a (non-isometric) 3D view added for visualization (produced by Dr. Shaoping Bai, currently Professor at Aalborg University, DK)

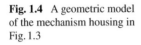

Fig. 1.4 A geometric model of the mechanism housing in Fig. 1.3

A *manufacturing* or *construction drawing* is, as a rule, a 2D display of the relevant dimensions and other information—e.g., tolerances and materials—required to produce parts or full design solutions. Sometimes, an axonometric view of the part in question is added for visualization purposes. Illustrated in Fig. 1.3 is a manufacturing drawing that shows one part of the mechanism housing sketched in Fig. 1.1.

A geometric model of the design solution is a more realistic representation of the same object, intended not only for visualization, but also for the calculation of the various geometric properties—volume, footprint area, centroid location, moment-of-inertia tensor—and mechanical behavior of the object. The latter can include stress and vibration analyses by means of CAE, an acronym for *Computer-Aided Engineering*. A geometric model of the same housing, with all its parts assembled, although not including the mechanism it is intended to hold, is illustrated in Fig. 1.4.

1.2 Back to Basics: Coordinate Space

1.2.1 The Cartesian Coordinate System

In order to locate points, lines, planes, or other geometric objects in space, the positions—and/or their orientation, as the case may be—of these objects must be known with respect to some reference frame. Generally, we use the *Cartesian* coordinate system to allow the position and orientation of geometric objects to be referenced relative to a selected frame.

As illustrated in Fig. 1.5, a two-dimensional coordinate system establishes an origin at the intersection of two mutually perpendicular axes, conventionally labeled X (horizontal) and Y (vertical). The origin is assigned the coordinate values (0, 0).

Using this coordinate system, we are able to construct a multitude of geometric objects by specifying the coordinates of the vertices and connecting them together with lines to form edges. An example of this is the rectangle shown in Fig. 1.6.

However, even though a 2D system may be useful for a variety of purposes, most real-world applications require a third dimension. The 2D Cartesian plane can readily be extended to allow for the inclusion of 3D points. In a 3D coordinate system, the

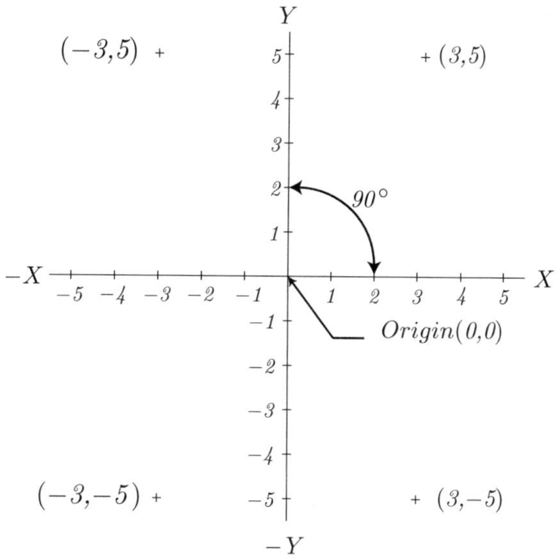

Fig. 1.5 A 2D coordinate system

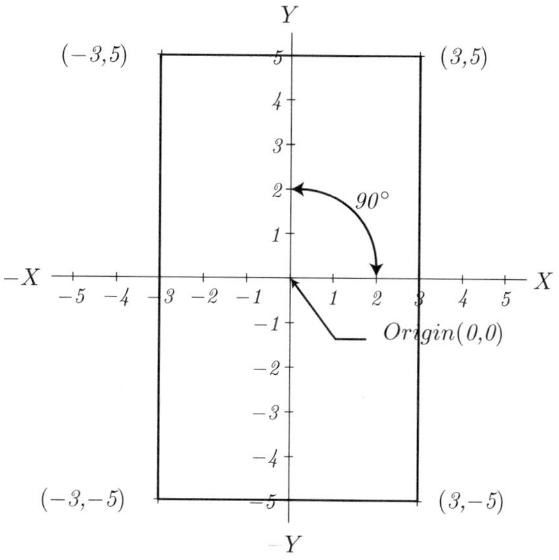

Fig. 1.6 Creating a rectangle

origin is established at the point where three mutually perpendicular axes—X, Y, and Z—intersect. The origin is assigned the coordinate values (0, 0, 0), as illustrated in Fig. 1.7.

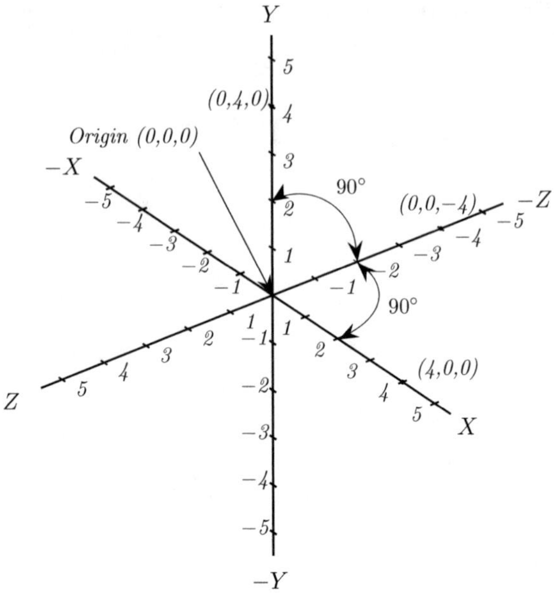

Fig. 1.7 3D coordinate system

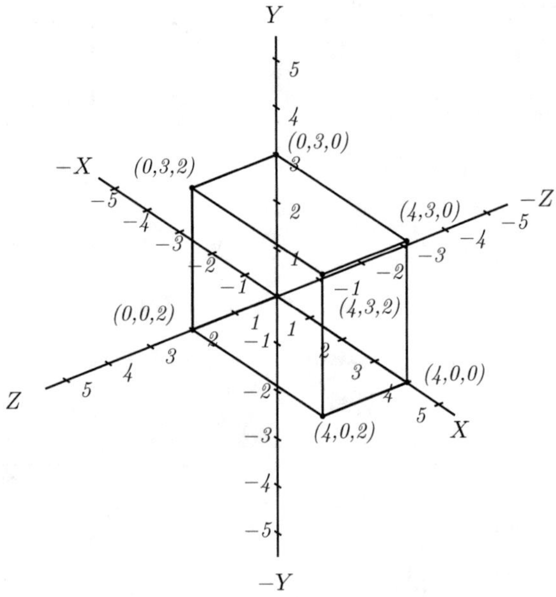

Fig. 1.8 Creating a rectangular parallelepiped

Similar to the 2D rectangle that was constructed, a rectangular parallelepiped is created using the 3D coordinate system by establishing coordinate values for each vertex, and then inserting the appropriate edges, as shown in Fig. 1.8.

This coordinate system is used in multiview drawings and 3D modeling, using CAD tools.

1.2.2 Right-Hand Rule

The right-hand rule is used to determine the positive direction of the axes; it defines the X-, Y-, and Z-axes as well as the positive and negative directions of rotation on each axis.

The simplest way to remember the right-hand rule is to first make a fist with your right hand—hence the rule name—ensuring that your thumb points upward. The direction in which your thumb points is the positive direction of the Z-axis. Now straighten your index finger so that it points straight at right angles with your thumb, thereby defining the X-axis. Finally, straighten your middle finger at right angles with both your thumb and your index finger, thereby defining the Y-axis.

It is noteworthy that, opposite to the right-hand rule, there exists a left-hand rule, which defines all of the previous axes in the same way as the right-hand rule, only *reflected*. While the left-hand rule is used in some situations to describe the coordinate axes, in this book we will only employ the conventional right-hand rule for the sake of consistency, simplicity, and while following established convention.

1.2.3 Types of Coordinate Systems

1.2.3.1 Polar Coordinates

Polar coordinates are used to locate points in the plane; they specify a distance ρ from a prescribed point, denoted the origin $O(0, 0)$, and an angle θ from a prescribed line, usually denoted the X-axis. Figure 1.9 shows a point P in the XY-plane, a distance ρ from the origin O, segment $\overline{OP}$ making an angle of θ with the X-axis, measured ccw[1] in the positive direction. Polar coordinates are commonly used by CAD systems to locate points because of their inherent simplicity.

The relations between polar and Cartesian coordinates are straightforward:

$$\rho = \sqrt{x^2 + y^2}, \quad \theta = \arctan(x, y) \tag{1.1a}$$

$$x = r \cos\theta, \quad y = r \sin\theta \tag{1.1b}$$

where $\arctan(x, y)$ is the *single-valued* arctan function, which returns the unique value of θ whose tangent is x/y, the correct quadrant being indicated by the sign of each x and y.

[1] Usual abbreviation for "counterclockwise"

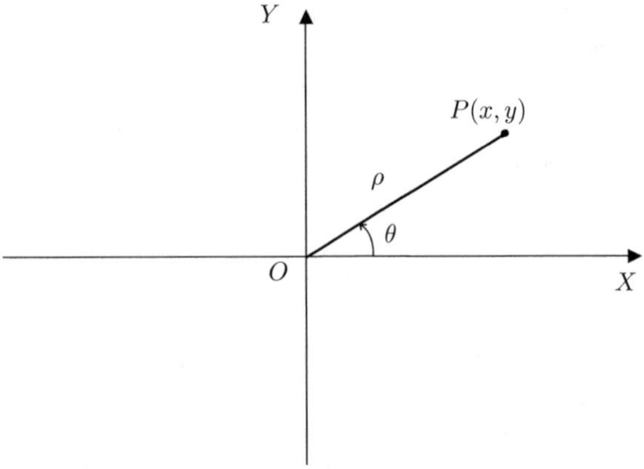

Fig. 1.9 Polar coordinates

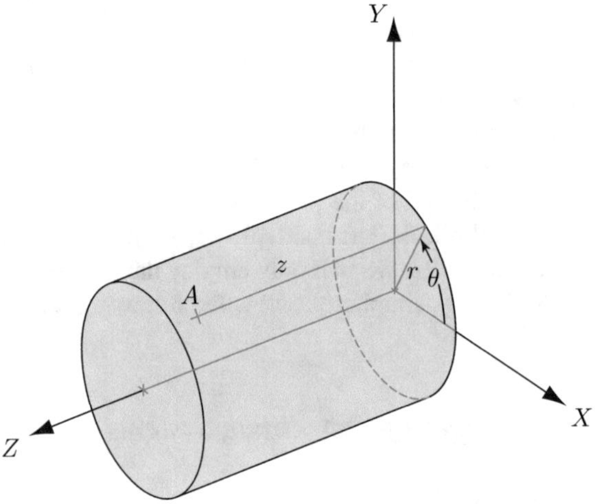

Fig. 1.10 Cylindrical coordinates

1.2.3.2 Cylindrical Coordinates

Cylindrical coordinates locate a point on the surface of a cylinder by specifying a distance from the origin and an angle from the X-axis in the XY-plane, besides the distance in the direction of Z. In Fig. 1.10, for example, point A is a distance z from the XY-plane, a distance r from the origin as measured on a line that makes an angle θ with the X-axis and lying in the XY-plane. Notice that the radius of the cylinder corresponds to the value of the cylindrical coordinate in question, with the point located on the surface of the cylinder itself.

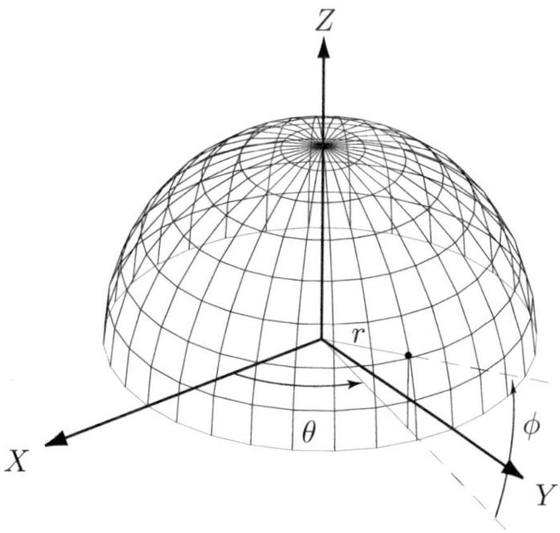

Fig. 1.11 Spherical coordinates

In general, cylindrical coordinates are used in designing axially symmetric shapes. The relations between cylindrical and Cartesian coordinates are given below:

$$r = \sqrt{x^2 + y^2}, \quad \theta = \arctan(x, y), \quad z = z \tag{1.2a}$$

$$x = r\cos\theta, \quad y = r\sin\theta, \quad z = z \tag{1.2b}$$

1.2.3.3 Spherical Coordinates

Spherical coordinates locate a point P on the surface of a sphere, as depicted in Fig. 1.11, by means of three *spherical coordinates*: the radius r of the sphere on which P lies; the *longitude* of the point, measured by angle θ; and the *latitude* of the point, measured by angle ϕ. In the figure, apparently, the X–Y plane defines the *equator*, the X–Z plane the *Greenwich meridian*.

The relations between Cartesian and spherical coordinates are readily derived from Fig. 1.11:

$$r = \sqrt{x^2 + y^2 + z^2}, \quad \phi = \arcsin\left(\frac{z}{r}\right), \quad \theta = \arccos\left(\frac{x}{\sqrt{x^2 + y^2}}\right) \tag{1.3a}$$

$$x = r\cos\phi\cos\theta, \quad y = r\cos\phi\sin\theta, z = r\sin\phi \tag{1.3b}$$

the ranges of ϕ and θ being

$$-\frac{\pi}{2} \le \phi \le \frac{\pi}{2}, \quad 0 \le \theta \le 2\pi \tag{1.3c}$$

That is, the sign of ϕ is the same as that of z, while the quadrant of θ is uniquely determined from the signs of x and y.

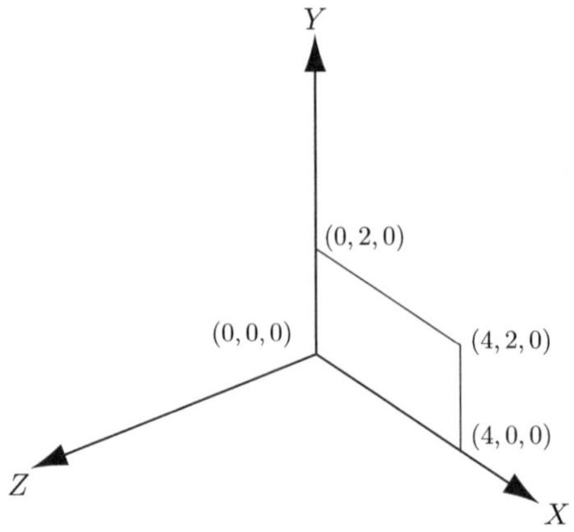

Fig. 1.12 Absolute coordinates

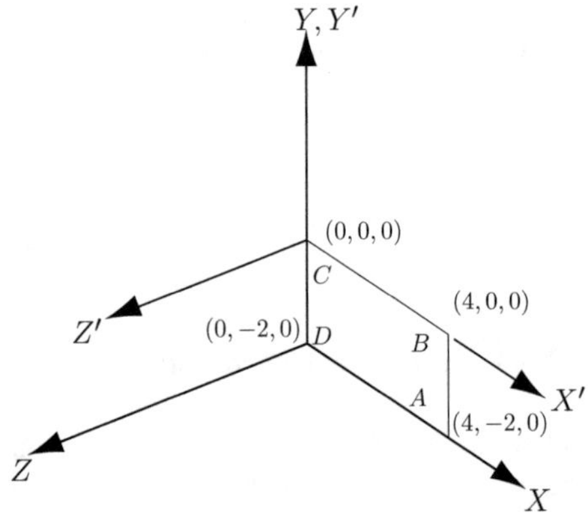

Fig. 1.13 World coordinates (x, y, x) versus relative coordinates (x', y', z')

1.2.3.4 Absolute/Relative/World Coordinate Systems

As illustrated in Fig. 1.12, *absolute coordinates* always refer to the origin $(0, 0, 0)$.

Relative coordinates are always referenced to a previously defined location and are sometimes referred to as Δ-coordinates, as pertaining to the $\{X', \ Y', \ Z'\}$ frame, and shown in Fig. 1.13. Thus, you can have several different coordinate systems within one larger coordinate system; each object may have its own local (relative) system, but each local system is referenced to an encompassing "world" coordinate system.

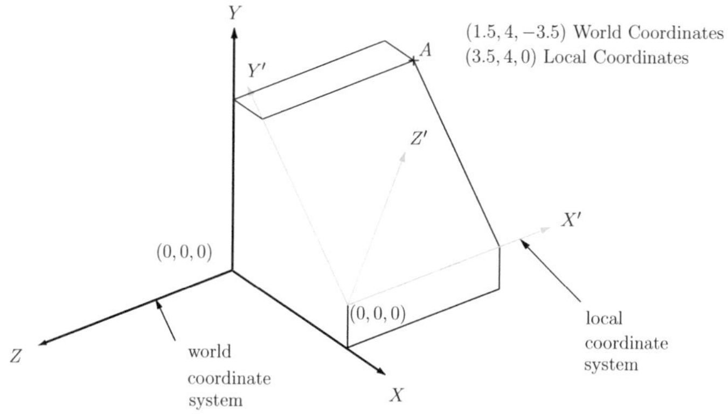

Fig. 1.14 World and local coordinates

As illustrated in Fig. 1.14, the *world coordinate system* uses a set of three numbers (x, y, z) located on three mutually perpendicular axes and measured from the origin $(0, 0, 0)$; the local coordinate system is a moving system that can be positioned anywhere in 3D space to assist in the construction of geometric objects.

1.2.4 Homogeneous Coordinates

One of the many purposes of *homogeneous coordinates* is to capture the concept of infinity. In the Euclidean coordinate system, infinity is something that does not exist. Mathematicians have discovered that many geometric concepts and computations can be greatly simplified if the concept of infinity is used. This will become apparent when we move to curve and surface design. Without the use of a homogeneous-coordinate system, it would be difficult to design certain classes of frequently used curves and surfaces in computer graphics and computer-aided design software, as well as perform transformations on these curves and surfaces. The concept of *point at infinity* is fundamental in geometry, but, as said above, difficult to represent with the aid of Euclidean geometry. *Projective geometry* comes to the rescue. In this concept, a point at infinity can be regarded as one whose distance from the observer is so large that its distance does not matter, only its *line of sight*, as in the case of stars, as far as the casual stargazer is concerned.

Any point P in the XY-plane has two coordinates. Adding a third component whose value is 1 to the coordinates of this point leads to the corresponding homogeneous coordinates. Thus, the homogeneous coordinates of any point P in the said plane, represented in the three-dimensional array $\boldsymbol{p}$, are

$$\boldsymbol{p} = \begin{bmatrix} x \\ y \\ 1 \end{bmatrix} \tag{1.4}$$

The distance d of P from the origin is known to be $d = \sqrt{x^2 + y^2}$. If we factor d out from the $\boldsymbol{p}$ array of Eq. (1.4), we obtain

$$\boldsymbol{p} = d \begin{bmatrix} \cos\theta \\ \sin\theta \\ 1/d \end{bmatrix}, \quad \theta = \cos^{-1}\left(\frac{y}{x}\right) \tag{1.5}$$

A point P_∞ lying at infinity, with a line of sight making an angle θ with the X-axis, has the homogeneous coordinates stored in the array $\boldsymbol{p}_\infty$ given below[2]:

$$\boldsymbol{p}_\infty = \begin{bmatrix} \cos\theta \\ \sin\theta \\ 0 \end{bmatrix} \tag{1.6}$$

which results from the array of Eq. (1.5) upon letting $d \to \infty$.

Similarly, the homogeneous coordinates of a point in 3D are defined as

$$\boldsymbol{p} = \begin{bmatrix} x \\ y \\ z \\ 1 \end{bmatrix} \tag{1.7}$$

Likewise, the homogeneous coordinates of a point P_∞ lying at infinity in 3D space, with a line of sight of direction cosines (λ, μ, ν), as formally defined in Eq. (1.19), are stored in an array $\boldsymbol{p}_\infty$ given as

$$\boldsymbol{p}_\infty = \begin{bmatrix} \lambda \\ \mu \\ \nu \\ 0 \end{bmatrix} \tag{1.8}$$

Homogeneous coordinates have been traditionally used instead of ordinary Cartesian coordinates in computer graphics and geometric modeling. The representation of points in homogeneous coordinates provides a unified approach to the description of geometric transformations and allows these transformations to be represented as simple matrix operations.

[2]Only the array is of interest, its coefficient, d in Eq. (1.5), being left out.

We will use homogeneous coordinates in Chap. 4 when we will study *affine transformations*. As we will see, homogenous coordinates simplify (and in many cases, make possible) the mathematics needed to represent the desired transformations and manipulations.

1.3 Vectors

1.3.1 Notation

Throughout this book, we use boldface fonts to indicate vectors ($\boldsymbol{a}$) and matrices ($\boldsymbol{R}$), with uppercase letters reserved for matrices and lowercases for vectors. Additionally, calligraphic literals ($\mathcal{C}$) are reserved for sets of points or other objects.

1.3.2 Definition

A vector is a mathematical entity that has[3]

- A direction,
- An orientation, and
- A norm (or magnitude).

A vector has a tail and a head, which determines its orientation. Conventionally, the head of a vector is indicated by an arrow, and the tail by a *thick* point. These two are joined together by a line to form the vector representation.

We now introduce the *unit vectors* $\boldsymbol{i}$, $\boldsymbol{j}$, and $\boldsymbol{k}$: vector $\boldsymbol{i}$ is parallel to the X-axis, $\boldsymbol{j}$ parallel to the Y-axis, and $\boldsymbol{k}$ parallel to the Z-axis. These vectors are given in component form as *column arrays* in displayed form; when included in lines of text, as row arrays, with a superscript T to indicate transposition. Thus,

$$\boldsymbol{i} = \begin{bmatrix} 1 \\ 0 \\ 0 \end{bmatrix} \quad \boldsymbol{j} = \begin{bmatrix} 0 \\ 1 \\ 0 \end{bmatrix} \quad \boldsymbol{k} = \begin{bmatrix} 0 \\ 0 \\ 1 \end{bmatrix} \tag{1.9}$$

Because we can multiply a vector by some scalar quantity that changes its magnitude but not its direction, we can express any given vector $\boldsymbol{v}$ as follows:

$$\boldsymbol{v} = v_x\boldsymbol{i} + v_y\boldsymbol{j} + v_z\boldsymbol{k} \tag{1.10}$$

[3]In the book, we are interested only in two- and three-dimensional vectors, hence our limited definition, which doesn't consider n-dimensional vectors common in linear algebra, or even ∞-dimensional vectors, proper of functional analysis.

This follows from $\boldsymbol{v} = \boldsymbol{v}_x + \boldsymbol{v}_y + \boldsymbol{v}_z$, where $\boldsymbol{v}_x = v_x\boldsymbol{i}$, $\boldsymbol{v}_y = v_y\boldsymbol{j}$, and $\boldsymbol{v}_z = v_z\boldsymbol{k}$. The same vector $\boldsymbol{v}$ can be described as a *column array*:

$$\boldsymbol{v} = \begin{bmatrix} v_x \\ v_y \\ v_z \end{bmatrix} \tag{1.11}$$

where $v_x, v_y, and v_z$ are the *components* of $\boldsymbol{v}$. The components may be negative, depending on the direction of the vector.

1.3.3 Basic Properties

1.3.3.1 Magnitude of a Vector

The magnitude of a vector, also known as the *Euclidean vector norm*, is non-negative and vanishes only when the vector itself does. The magnitude of $\boldsymbol{v}$ is thus a non-negative scalar quantity, denoted by $\|\boldsymbol{v}\|$ and given by

$$\|\boldsymbol{v}\| = \sqrt{v_x^2 + v_y^2 + v_z^2} \tag{1.12}$$

which is a simple application of the Pythagorean theorem used to find the length of the diagonal of a parallelepiped of sides with lengths v_x, v_y, v_z, such as the one shown in Fig. 1.8. Hence,

$$\|\boldsymbol{v}\|^2 = v_x^2 + v_y^2 + v_z^2 \tag{1.13}$$

1.3.3.2 Unit Vector

We define a unit vector as any vector whose magnitude is equal to unity, regardless of its direction. As we saw, $\boldsymbol{i}$, $\boldsymbol{j}$, and $\boldsymbol{k}$ are special cases of unit vectors, with specific directions assigned to them. An arbitrary unit vector in the direction of $\boldsymbol{v}$ can be obtained as follows:

$$\boldsymbol{w} = \frac{\boldsymbol{v}}{\|\boldsymbol{v}\|} \tag{1.14}$$

with

$$\|\boldsymbol{w}\| = 1 \tag{1.15}$$

A unit vector can be also written in the form

$$\boldsymbol{w} = \begin{bmatrix} v_x/\|\boldsymbol{v}\| \\ v_y/\|\boldsymbol{v}\| \\ v_z/|\boldsymbol{v}\| \end{bmatrix} \equiv \frac{1}{\|\boldsymbol{v}\|} \begin{bmatrix} v_x \\ v_y \\ v_z \end{bmatrix} \tag{1.16}$$

We can make this more concise with the definitions:

$$w_x = \frac{v_x}{\|\boldsymbol{v}\|}, \; w_y = \frac{v_y}{\|\boldsymbol{v}\|}, \; w_z = \frac{v_z}{\|\boldsymbol{v}\|} \tag{1.17}$$

so that

$$\boldsymbol{w} = \begin{bmatrix} w_x \\ w_y \\ w_z \end{bmatrix} \tag{1.18}$$

Note that if α, β, and γ are the angles between $\boldsymbol{v}$ and the X-, Y-, and Z-axes, respectively, then

$$w_x = \frac{v_x}{\|\boldsymbol{v}\|} = \cos\alpha \qquad w_y = \frac{v_y}{\|\boldsymbol{v}\|} = \cos\beta \qquad w_z = \frac{v_z}{\|\boldsymbol{v}\|} = \cos\gamma \tag{1.19}$$

which indicates that w_x, w_y, and w_z are the *direction cosines* of $\boldsymbol{v}$, whose squares add up to unity.

1.3.3.3 Multiplication by a Scalar

Multiplying any vector $\boldsymbol{v}$ by a scalar k produces a new vector $k\boldsymbol{v}$:

$$k\boldsymbol{v} = \begin{bmatrix} kv_x \\ kv_y \\ kv_z \end{bmatrix} \tag{1.20}$$

If k is positive, then $\boldsymbol{v}$ and $k\boldsymbol{v}$ are in the same direction; if k is negative, then $\boldsymbol{v}$ and $k\boldsymbol{v}$ are in opposite directions. The magnitude of $k\boldsymbol{v}$ is

$$\|k\boldsymbol{v}\| = \sqrt{k^2v_x^2 + k^2v_y^2 + k^2v_z^2} \tag{1.21}$$

so that

$$\|k\boldsymbol{v}\| = |k|\|\boldsymbol{v}\| \tag{1.22}$$

1.3.3.4 Vector Addition

Given $\boldsymbol{a} = [a_x \; a_y \; a_z]^T$ and $\boldsymbol{b} = [b_x \; b_y \; b_z]^T$, the sum of these two vectors is defined as

$$\boldsymbol{a} + \boldsymbol{b} = \begin{bmatrix} a_x + b_x \\ a_y + b_y \\ a_z + b_z \end{bmatrix} \tag{1.23}$$

Given vectors $\boldsymbol{a}, \boldsymbol{b}, \boldsymbol{c}$ and scalars k and l, vector addition and scalar multiplication obey the properties below:

1. $\boldsymbol{a} + \boldsymbol{b} = \boldsymbol{b} + \boldsymbol{a}$,
2. $\boldsymbol{a} + (\boldsymbol{b} + \boldsymbol{c}) = (\boldsymbol{a} + \boldsymbol{b}) + \boldsymbol{c}$,
3. $k(l\boldsymbol{a}) = kl\boldsymbol{a}$,
4. $(k + l)\boldsymbol{a} = k\boldsymbol{a} + l\boldsymbol{a}$, and
5. $k(\boldsymbol{a} + \boldsymbol{b}) = k\boldsymbol{a} + k\boldsymbol{b}$.

1.3.4 Scalar Product

The scalar product, also known as the *dot product*, of two vectors $\boldsymbol{a}$ and $\boldsymbol{b}$ is the sum of the products of their corresponding components:

$$\boldsymbol{a} \cdot \boldsymbol{b} = a_x b_x + a_y b_y + a_z b_z \tag{1.24}$$

which returns a scalar quantity, not another vector. An alternative form representing the scalar product is

$$\boldsymbol{a}^T \boldsymbol{b} = a_x b_x + a_y b_y + a_z b_z \tag{1.25}$$

with superscript T indicating *transposition*, i.e., $\boldsymbol{a}^T$ is the transpose of $\boldsymbol{a}$. Since a vector array was defined in Sect. 1.2.4 as a *column array*, $\boldsymbol{a}^T$ is a *row array*.

In addition, we can readily show that the scalar product is commutative, meaning that:

$$\boldsymbol{a} \cdot \boldsymbol{b} = \boldsymbol{b} \cdot \boldsymbol{a} \tag{1.26}$$

Alternatively, the scalar product may be calculated using the angle θ between the vectors $\boldsymbol{a}$ and $\boldsymbol{b}$:

$$\boldsymbol{a} \cdot \boldsymbol{b} = \|\boldsymbol{a}\| \|\boldsymbol{b}\| \cos\theta \tag{1.27}$$

Moreover, the following two statements are equivalent: for any $\boldsymbol{a}, \boldsymbol{b} \neq \boldsymbol{0}$,

$$\boldsymbol{a} \cdot \boldsymbol{b} = 0 \Longleftrightarrow \boldsymbol{a} \text{ and } \boldsymbol{b} \text{ are perpendicular.} \tag{1.28}$$

In summary, the scalar product has the following properties:

1. $\boldsymbol{a} \cdot \boldsymbol{b} = \|\boldsymbol{a}\| \|\boldsymbol{b}\| \cos\theta$, where θ is the angle between $\boldsymbol{a}$ and $\boldsymbol{b}$;
2. $\boldsymbol{a} \cdot \boldsymbol{a} = \|\boldsymbol{a}\|^2$;
3. $\boldsymbol{a} \cdot \boldsymbol{b} = \boldsymbol{b} \cdot \boldsymbol{a}$, *commutativity*;
4. $\boldsymbol{a} \cdot (\boldsymbol{b} + \boldsymbol{c}) = \boldsymbol{a} \cdot \boldsymbol{b} + \boldsymbol{a} \cdot \boldsymbol{c}$, *distributivity*;
5. $(k\boldsymbol{a}) \cdot \boldsymbol{b} = \boldsymbol{a} \cdot (k\boldsymbol{b}) = k(\boldsymbol{a} \cdot \boldsymbol{b})$, *associativity*; and
6. $\boldsymbol{a}$ is perpendicular to $\boldsymbol{b} \Longleftrightarrow \boldsymbol{a} \cdot \boldsymbol{b} = 0$.

1.3.5 Inequalities

In connection with vector norms, there are two important inequalities that arise, as given below:

- Cauchy–Schwartz inequality:

$$(\boldsymbol{v} \cdot \boldsymbol{w})^2 \leq \|\boldsymbol{v}\|^2 \|\boldsymbol{w}\|^2 \tag{1.29}$$

- Triangle inequality:

$$\|\boldsymbol{v} + \boldsymbol{w}\| \leq \|\boldsymbol{v}\| + \|\boldsymbol{w}\| \tag{1.30}$$

These properties are quite useful for the derivation of other identities and inequalities, and are fundamental to the understanding and application of vectors in computer graphics.

1.4 Matrices

1.4.1 Definition

A matrix is a rectangular array of numbers arranged in m rows and n columns, namely,

$$\boldsymbol{A} = \begin{bmatrix} a_{11} & a_{12} & a_{13} & \dots & a_{1n} \\ a_{21} & a_{22} & a_{23} & \dots & a_{2n} \\ \vdots & \vdots & \vdots & \ddots & \vdots \\ a_{m1} & a_{m2} & a_{m3} & \dots & a_{mn} \end{bmatrix} \tag{1.31}$$

1.4.2 Special Matrices

- Square matrix:
  A square matrix has an equal number of rows and columns ($m = n$), e.g.,

$$\boldsymbol{A} = \begin{bmatrix} a_{11} & a_{12} & a_{13} \\ a_{21} & a_{22} & a_{23} \\ a_{31} & a_{32} & a_{33} \end{bmatrix} \tag{1.32}$$

- Row matrix:
  A row matrix has one row:

$$\boldsymbol{A} = \begin{bmatrix} a_{11} & a_{12} & a_{13} \end{bmatrix} = \boldsymbol{a}^T \tag{1.33}$$

That is, a row matrix is a *transposed* vector.

- Column matrix:
  A column matrix has one column. In our choice of notation, this column matrix becomes a vector array, namely,

$$\boldsymbol{b} = \begin{bmatrix} b_{11} \\ b_{21} \\ b_{31} \end{bmatrix} \tag{1.34}$$

  Note the use of a bold *lowercase* font in Eqs. (1.33) and (1.34), indicating that this array is a vector.
- Diagonal matrix:
  A square matrix that has zero entries everywhere except on its main diagonal is termed *diagonal*:

$$\boldsymbol{A} = \begin{bmatrix} a_{11} & 0 & 0 & 0 & \dots & 0 \\ 0 & a_{22} & 0 & 0 & \dots & \vdots \\ 0 & 0 & a_{33} & 0 & \dots & \vdots \\ 0 & 0 & 0 & a_{44} & \dots & \vdots \\ \vdots & \vdots & \vdots & \vdots & \ddots & 0 \\ 0 & \dots & \dots & \dots & 0 & a_{mm} \end{bmatrix} \tag{1.35}$$

  Diagonal matrices satisfy: $a_{ij} = 0$ if $i \neq j$.
- Identity matrix:
  This is a special diagonal matrix with unit elements on the main diagonal, and zero elsewhere. This matrix is denoted by the symbol **1**. For example, the 3×3 identity matrix is defined as

$$\mathbf{1} = \begin{bmatrix} 1 & 0 & 0 \\ 0 & 1 & 0 \\ 0 & 0 & 1 \end{bmatrix} \tag{1.36}$$

  The elements of **1** are sometimes represented by the *Kronecker delta*, namely,

$$\begin{aligned} \delta_{ij} &= 0 \quad \text{if } i \neq j \\ \delta_{ij} &= 1 \quad \text{if } i = j \end{aligned} \tag{1.37}$$

- Zero matrix:
  The $m \times n$ zero matrix has all its elements equal to zero. We will denote it by $\boldsymbol{O}$. For example, the 3×3 zero matrix is

$$\boldsymbol{O} = \begin{bmatrix} 0 & 0 & 0 \\ 0 & 0 & 0 \\ 0 & 0 & 0 \end{bmatrix} \tag{1.38}$$

- Symmetric matrix:
Symmetric matrices are symmetric about the main diagonal: $a_{ij} = a_{ji}$, i.e.,

$$\boldsymbol{A} = \begin{bmatrix} a_{11} & a_{12} & a_{13} \\ a_{12} & a_{22} & a_{23} \\ a_{13} & a_{23} & a_{33} \end{bmatrix} = \begin{bmatrix} a_{11} & a_{12} & a_{13} \\ \vdots & a_{22} & a_{23} \\ \text{symmetric} & \dots & a_{33} \end{bmatrix} \tag{1.39}$$

- Skew-symmetric matrix:
A matrix is skew-symmetric if $a_{ij} = -a_{ji}$, namely,

$$\boldsymbol{A} = \begin{bmatrix} 0 & a_{12} & a_{13} \\ -a_{12} & 0 & a_{23} \\ -a_{13} & -a_{23} & 0 \end{bmatrix} \tag{1.40}$$

Notice that the diagonal entries of a skew-symmetric matrix vanish necessarily.
- Triangular matrix: An *upper triangular* matrix has all its entries below the diagonal equal to zero, namely,

$$\boldsymbol{U} = \begin{bmatrix} a_{11} & a_{12} & \dots & a_{1n} \\ 0 & a_{22} & \dots & a_{2n} \\ \vdots & \vdots & \ddots & \vdots \\ 0 & 0 & \dots & a_{nn} \end{bmatrix} \tag{1.41}$$

A *lower triangular* matrix is defined correspondingly. Furthermore, while upper (lower) triangular matrices can be rectangular, *symmetric* and *skew-symmetric* matrices are necessarily square.

1.4.3 Properties

- Matrix equality: Let us assume that matrices $\boldsymbol{A}$ and $\boldsymbol{B}$ have the same numbers of rows and columns. Then,

$$\boldsymbol{A} = \boldsymbol{B} \Longleftrightarrow a_{ij} = b_{ij}, \text{ for all } i, j \tag{1.42}$$

- Matrix addition:
Adding two $m \times n$ matrices $\boldsymbol{A}$ and $\boldsymbol{B}$ produces a third $m \times n$ matrix $\boldsymbol{C}$, whose entries are equal to the sum of the corresponding entries of $\boldsymbol{A}$ and $\boldsymbol{B}$. Thus,

$$\boldsymbol{A} + \boldsymbol{B} = \boldsymbol{C} \tag{1.43}$$

and

$$a_{ij} + b_{ij} = c_{ij} \tag{1.44}$$

Of course, we can add or subtract two matrices if and only if they have the same numbers of rows and columns. Moreover, matrix addition is commutative, i.e., $\boldsymbol{A} + \boldsymbol{B} = \boldsymbol{B} + \boldsymbol{A}$.

- Matrix multiplication by a scalar:
  Multiplying a matrix $\boldsymbol{A}$ by a scalar k produces a new matrix $\boldsymbol{B}$ with the same number of rows and columns. Each entry of $\boldsymbol{B}$ is obtained by multiplying the corresponding entry of $\boldsymbol{A}$ by the scalar k:

$$k\boldsymbol{A} = \boldsymbol{B} \tag{1.45}$$

and

$$ka_{ij} = b_{ij} \tag{1.46}$$

- Matrix product:
  The product $\boldsymbol{AB}$ of two matrices is another matrix $\boldsymbol{C}$. This operation is possible if and only if the number of columns of the first matrix is equal to the number of rows of the second matrix. In general, the product of two matrices is not commutative:

$$\boldsymbol{AB} \neq \boldsymbol{BA} \tag{1.47}$$

The product $\boldsymbol{C}$ of two matrices $\boldsymbol{A}$ and $\boldsymbol{B}$ in terms of their entries is

$$\boldsymbol{C} = \boldsymbol{AB} \quad \Rightarrow \quad c_{ij} = a_{i1}b_{1j} + a_{i2}b_{2j} + \cdots + a_{in}b_{nj} \tag{1.48}$$

- Transpose matrix
  By interchanging the rows and columns of a matrix $\boldsymbol{A}$, we obtain its transpose $\boldsymbol{A}^T$, so that $a^T_{ij} = a_{ji}$, where a^T_{ij} is the (i, j) entry of the transpose of $\boldsymbol{A}$, i.e.,

$$\boldsymbol{A} = \begin{bmatrix} a_{11} & a_{12} \\ a_{21} & a_{22} \\ a_{31} & a_{32} \end{bmatrix} \quad \Rightarrow \quad \boldsymbol{A}^T = \begin{bmatrix} a_{11} & a_{21} & a_{31} \\ a_{12} & a_{22} & a_{32} \end{bmatrix} \tag{1.49}$$

1.4.4 The Vector Product

The vector product of two 3D vectors $\boldsymbol{a}$ and $\boldsymbol{b}$, also known as the *cross product*, is defined as

$$\boldsymbol{a} \times \boldsymbol{b} = (a_y b_z - a_z b_y)\boldsymbol{i} - (a_x b_z - a_z b_x)\boldsymbol{j} + (a_x b_y - a_y b_x)\boldsymbol{k} \tag{1.50}$$

In array form,

$$\boldsymbol{a} \times \boldsymbol{b} = \begin{bmatrix} a_y b_z - a_z b_y \\ a_z b_x - a_x b_z \\ a_x b_y - a_y b_x \end{bmatrix} \tag{1.51}$$

One *mnemonic* means to compute the vector product relies on the expansion of a determinant (cofactor expansion, as outlined in Sect. 1.4.6), namely,

$$\begin{aligned}\boldsymbol{a} \times \boldsymbol{b} &= \det \begin{bmatrix} \boldsymbol{i} & \boldsymbol{j} & \boldsymbol{k} \\ a_x & a_y & a_z \\ b_x & b_y & b_z \end{bmatrix} \\ &= (a_y b_z - a_z b_y)\boldsymbol{i} - (a_x b_z - a_z b_x)\boldsymbol{j} + (a_x b_y - a_y b_x)\boldsymbol{k} \end{aligned} \tag{1.52}$$

The properties of the vector product are given below:

- If $\boldsymbol{c} = \boldsymbol{a} \times \boldsymbol{b}$, then $\boldsymbol{c}$ is perpendicular to both $\boldsymbol{a}$ and $\boldsymbol{b}$. Consequently, $\boldsymbol{c}$ is also perpendicular to the plane defined by $\boldsymbol{a}$ and $\boldsymbol{b}$.
- The vector product is *skew-symmetric*: $\boldsymbol{a} \times \boldsymbol{b} = -(\boldsymbol{b} \times \boldsymbol{a})$.
- $\boldsymbol{a}$ and $\boldsymbol{b}$ are parallel $\Longleftrightarrow \boldsymbol{a} \times \boldsymbol{b} = \boldsymbol{0}$.

Also note that, if $\boldsymbol{c} = \boldsymbol{a} \times \boldsymbol{b}$ and θ denotes the angle from $\boldsymbol{a}$ to $\boldsymbol{b}$, in this direction, when the tails of the two vectors coincide, then

$$\boldsymbol{c} = (\|\boldsymbol{a}\|\|\boldsymbol{b}\| \sin\theta)\boldsymbol{n} \tag{1.53}$$

where $\boldsymbol{n}$ is the unit vector normal to both $\boldsymbol{a}$ and $\boldsymbol{b}$, so that $(\boldsymbol{a}, \boldsymbol{b}, \boldsymbol{n})$ is a right-hand triad.

The *geometric interpretation* of the vector product is now straightforward: As depicted in Fig. 1.15, the expression in parentheses in Eq. (1.53) is nothing but *twice* the area of the triangle AOB, i.e., the area of the parallelogram $OADB$.

1.4.5 The 2D Form of the Vector (Cross) Product

The vector product, or cross product, of two 3D vectors was defined in Sect. 1.4.4. This product exists only in three dimensions. However, in 2D geometry one is confronted frequently with the calculation of the cross product. To ease the solution of 2D geometric problems involving the cross product, we introduce below a *2D form of the cross product*.

Let $\boldsymbol{E}$ be an *orthogonal matrix* that rotates vectors in the plane through an angle of 90° counterclockwise (ccw), namely,

$$\boldsymbol{E} \equiv \begin{bmatrix} 0 & -1 \\ 1 & 0 \end{bmatrix} \tag{1.54}$$

With this definition, we can readily prove that

$$\boldsymbol{E}^T\boldsymbol{E} = \boldsymbol{E}\boldsymbol{E}^T = \boldsymbol{1} \tag{1.55}$$

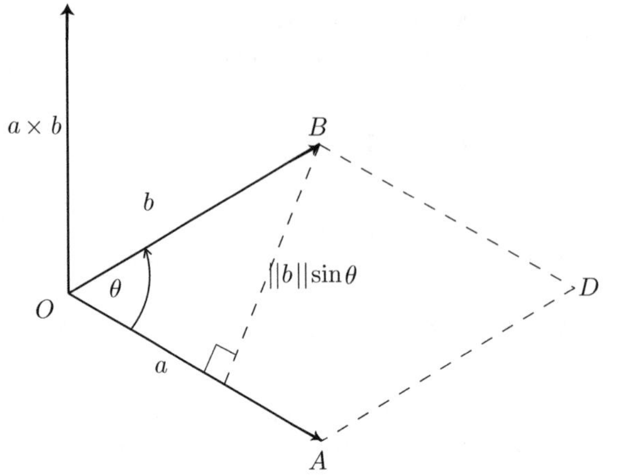

Fig. 1.15 Geometric interpretation of the vector product

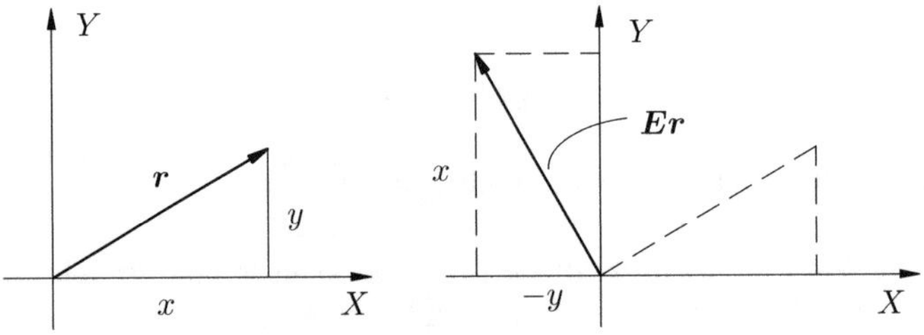

Fig. 1.16 Vector $\boldsymbol{r}$ and its image under $\boldsymbol{E}$

in which $\mathbf{1}$ is the 2×2 identity matrix. Moreover, note that $\boldsymbol{E}$ is *skew-symmetric*, i.e., $\boldsymbol{E} = -\boldsymbol{E}^T$, and hence $\boldsymbol{E}^2 = -\mathbf{1}$.

Also note that, given any vector $\boldsymbol{r} = \begin{bmatrix} x & y \end{bmatrix}^T$ in a plane Π, its image under $\boldsymbol{E}$ is given by

$$\boldsymbol{Er} = \begin{bmatrix} -y \\ x \end{bmatrix} \tag{1.56}$$

as illustrated in Fig. 1.16.

Now, let us compute the cross product $\boldsymbol{a} \times \boldsymbol{b}$, where $\boldsymbol{a} = A\boldsymbol{k}$, and $\boldsymbol{k}$ is a unit vector normal to Π, pointing toward the viewer, $\boldsymbol{a}$ thus being a 3D vector normal to Π, of magnitude $|A|$. Moreover, we assume that $\boldsymbol{b}$ lies in Π, its Z-component thus vanishing. The cross product of interest thus takes the form[4].

$$\boldsymbol{a} \times \boldsymbol{b} = \det \begin{bmatrix} \boldsymbol{i} & \boldsymbol{j} & \boldsymbol{k} \\ 0 & 0 & A \\ b_x & b_y & 0 \end{bmatrix} = -Ab_y\boldsymbol{i} + Ab_x\boldsymbol{j} \tag{1.57}$$

[4]"det" is abbreviation for *determinant*. A concept formally defined in Subsection 1.4.6.

where we have recalled that the unit vectors $\boldsymbol{i}$ and $\boldsymbol{j}$ are parallel to the X- and Y-axes, respectively. The two-dimensional form of the foregoing product, then, becomes

$$(\boldsymbol{a} \times \boldsymbol{b})_{2D} = A \begin{bmatrix} -b_y \\ b_x \end{bmatrix} \equiv A\boldsymbol{E}\overline{\boldsymbol{b}} \tag{1.58}$$

where we have recalled Eq. (1.56) and $\overline{\boldsymbol{b}}$ denotes the 2D version of the three-dimensional $\boldsymbol{b}$, i.e.,

$$\overline{\boldsymbol{b}} = \begin{bmatrix} b_x \\ b_y \end{bmatrix}$$

Likewise, the cross product $\boldsymbol{b} \times \boldsymbol{c}$, for both $\boldsymbol{b}$ and $\boldsymbol{c}$ in the plane Π, is a vector perpendicular to this plane, of *signed* magnitude,[5] $\|\boldsymbol{b}\|\|\boldsymbol{c}\| \sin(\boldsymbol{b}, \boldsymbol{c})$, where $(\boldsymbol{b}, \boldsymbol{c})$ denotes the angle between these two vectors, measured from $\boldsymbol{b}$ to $\boldsymbol{c}$.

Thus, if $\sin(\boldsymbol{b}, \boldsymbol{c})$ is positive, the cross-product vector points in the direction of $\boldsymbol{k}$; otherwise, in the direction of $-\boldsymbol{k}$.

More concretely, let $\boldsymbol{b}$ be defined as before, $\boldsymbol{c}$ being defined, in turn, as

$$\boldsymbol{c} \equiv \begin{bmatrix} c_x \\ c_y \\ 0 \end{bmatrix} \tag{1.59}$$

Hence,

$$\boldsymbol{b} \times \boldsymbol{c} = \det \begin{bmatrix} \boldsymbol{i} & \boldsymbol{j} & \boldsymbol{k} \\ b_x & b_y & 0 \\ c_x & c_y & 0 \end{bmatrix} = (b_x c_y - b_y c_x)\boldsymbol{k} \equiv C\boldsymbol{k} \tag{1.60}$$

where C is a real number that can be positive, negative, or even 0. Since we know the direction of $\boldsymbol{b} \times \boldsymbol{c}$, i.e., perpendicular to the plane Π, all we need is the quantity C above, which can be readily recognized as the dot product of the two-dimensional vectors $\boldsymbol{E}\overline{\boldsymbol{b}}$, as given in Eq. (1.58), and $\overline{\boldsymbol{c}}$, the 2D counterpart of $\boldsymbol{c}$, i.e.,

$$C = \begin{bmatrix} c_x & c_y \end{bmatrix}^T \begin{bmatrix} -b_y \\ b_x \end{bmatrix} = \overline{\boldsymbol{c}}^T \boldsymbol{E}\overline{\boldsymbol{b}} \equiv (\boldsymbol{E}\overline{\boldsymbol{b}})^T \overline{\boldsymbol{c}} = -\overline{\boldsymbol{b}}^T \boldsymbol{E}\overline{\boldsymbol{c}} \tag{1.61}$$

Therefore, the sign of C depends on whether the 3D cross product of Eq. (1.60) points toward the reader or not. C vanishes, of course, if the two factors, $\overline{\boldsymbol{b}}$ and $\overline{\boldsymbol{c}}$, are parallel.

Next we recall a concept of analytic geometry: a *directed segment* $\overrightarrow{\mathcal{L}}$ is a line $\mathcal{L}$ with a direction. In Fig. 1.17, two directed segments $\overrightarrow{\mathcal{L}}_1$ and $\overrightarrow{\mathcal{L}}_2$ are shown, each given by a line $\mathcal{L}_i$ and a unit vector $\mathbf{e}_i$, for $i = 1, 2$.

---

[5]The signed magnitude of a vector is a real number, positive, negative, or zero, whose absolute value is identical to the magnitude of the vector.

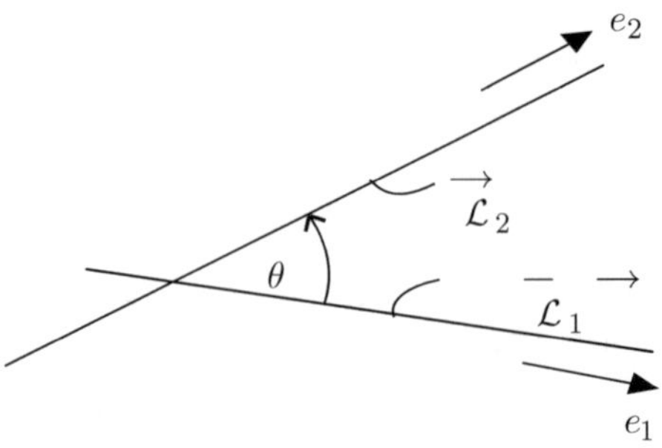

Fig. 1.17 The angle between two directed segments, measured from $\vec{\mathcal{L}}_1$ to $\vec{\mathcal{L}}_2$

Now we introduce a practical application of the foregoing concepts in solving a recurrent problem of planar geometry:

Problem Given two directed segments $\vec{\mathcal{L}}_1$ and $\vec{\mathcal{L}}_2$, find the angle θ, for $0 \leq \theta \leq 2\pi$, that $\vec{\mathcal{L}}_2$ makes with $\vec{\mathcal{L}}_1$, while measuring θ ccw.

Solution Let $\boldsymbol{e}_1$ and $\boldsymbol{e}_2$ be 2D unit vectors indicating the direction of segments $\vec{\mathcal{L}}_1$ and $\vec{\mathcal{L}}_2$, respectively. Obviously, $\cos\theta$ can be derived from the scalar product P1 of $\boldsymbol{e}_1$ and $\boldsymbol{e}_2$, namely,

$$\mathrm{P1} = \cos\theta = \boldsymbol{e}_1^T\boldsymbol{e}_2 \quad \text{or} \quad \boldsymbol{e}_1 \cdot \boldsymbol{e}_2 \tag{1.62}$$

However, the foregoing value does not determine uniquely θ, for, if $\mathrm{P1} > 0$, then θ may lie in either the first or the fourth quadrant, thus leaving us with an ambiguity. Ditto if $\mathrm{P1} < 0$, in which case θ may lie in either the second or the third quadrant.

To destroy the ambiguity, we need $\sin\theta$, which can be derived from the vector product of $\boldsymbol{e}_1$ and $\boldsymbol{e}_2$ when regarded as 3D vectors, as per Eq. (1.53). In light of the 2D form of the vector product, however, we need not work out of the plane of the two given lines. Indeed, from Eqs. (1.53), (1.60), and (1.61), we can write

$$P2 = \sin\theta = (\boldsymbol{E}\boldsymbol{e}_1)^T\boldsymbol{e}_2 \quad \text{or} \quad (\boldsymbol{E}\boldsymbol{e}_1) \cdot \boldsymbol{e}_2 \tag{1.63}$$

Obviously, if $\mathrm{P2} > 0$, then θ may lie in either the first or the second quadrant, which leaves us with an ambiguity. Ditto if $\mathrm{P2} < 0$, in which case θ may lie in either the third or the fourth quadrant.

While each of P1 and P2 does not determine *univocally* angle θ individually, both do. In fact, we can draw the rules below:

1. If P1 > 0 and P2 > 0, then θ lies in the first quadrant;
2. If P1 < 0 and P2 > 0, then θ lies in the second quadrant;
3. If P1 < 0 and P2 < 0, then θ lies in the third quadrant;
4. If P1 > 0 and P2 < 0, then θ lies in the fourth quadrant.

1.4.6 Determinants

The *determinant* is a quantity associated with an arbitrary $n \times n$ square matrix (note that the number of rows and columns must be identical). A general definition of what a determinant actually represents is rather cumbersome; luckily, we do not need it here. We can define the determinant of a $n \times n$ matrix starting with the simplest case, i.e., $n = 2$, then $n = 3$, and hence, by induction, derive a *procedure* to compute the determinant for any arbitrary value of n.

As a matter of fact, the interest in the determinant is rather theoretical; its actual computation, which is extremely costly in terms of *floating-point operations*, or *flops*, is seldom needed. The relevance of the concept lies in that the value of the determinant indicates whether or not the matrix is *singular*; if it is so, then the *solution* to the system of linear equations associated with the matrix coefficient is not unique. We are interested, in this book, only in the case in which the determinant is *non-zero*, in which case the associated system of linear equations admits one solution, which is *unique*. In this case, the matrix is said to be *invertible*, *regular* or, most commonly, *non-singular*.

A 2×2 matrix $\boldsymbol{A}$ can be *partitioned* either column-wise or row-wise, as shown below:

$$\boldsymbol{A} \equiv \begin{bmatrix} \boldsymbol{a} & \boldsymbol{b} \end{bmatrix} \equiv \begin{bmatrix} \boldsymbol{c}^T \\ \boldsymbol{d}^T \end{bmatrix} \tag{1.64}$$

where $\boldsymbol{a}$, $\boldsymbol{b}$, $\boldsymbol{c}$, and $\boldsymbol{d}$ are all two-dimensional column vectors. Furthermore, we recall the definition of $\boldsymbol{E}$ as seen in Sect. 1.4.5. If the components of $\boldsymbol{a}$ and $\boldsymbol{b}$ are given as $\boldsymbol{a} = \begin{bmatrix} a_x & a_y \end{bmatrix}^T$ and $\boldsymbol{b} = \begin{bmatrix} b_x & b_y \end{bmatrix}^T$, then the determinant of $\boldsymbol{A}$ is defined as

$$\det(\boldsymbol{A}) = \det \begin{bmatrix} a_x & b_x \\ a_y & b_y \end{bmatrix} = a_x b_y - a_y b_x \tag{1.65}$$

which can be readily cast in the form

$$\det(\boldsymbol{A}) = \begin{bmatrix} a_x & a_y \end{bmatrix} \begin{bmatrix} b_y \\ -b_x \end{bmatrix} \equiv \begin{bmatrix} a_x & a_y \end{bmatrix} \underbrace{\left(- \begin{bmatrix} -b_y \\ b_x \end{bmatrix} \right)}_{-\boldsymbol{E}\boldsymbol{b}} \tag{1.66}$$

the first array of the foregoing product being $\boldsymbol{a}^T$, the second $-\boldsymbol{E}\boldsymbol{b}$.

In summary, then

$$\det(\boldsymbol{A}) = -\boldsymbol{a}^T\boldsymbol{E}\boldsymbol{b} = \boldsymbol{b}^T\boldsymbol{E}\boldsymbol{a} \tag{1.67}$$

where we have recalled that (i) from Eq. (1.26), the scalar product is *commutative* and (ii) from Eq. (1.54), that $\boldsymbol{E}$ is *skew-symmetric*.

A property of the determinant follows from its definition given in Eq. (1.65):

$$\det(\boldsymbol{A}) = \det(\boldsymbol{A}^T) \tag{1.68}$$

If we recall the column-wise partitioning of $\boldsymbol{A}$, we can readily conclude that

$$\boldsymbol{A}^T = \begin{bmatrix}\boldsymbol{c} & \boldsymbol{d}\end{bmatrix} \tag{1.69}$$

thereby obtaining an alternative expression for $\det(\boldsymbol{A})$:

$$\det(\boldsymbol{A}) = \det(\boldsymbol{A}^T) = -\boldsymbol{c}^T\boldsymbol{E}\boldsymbol{d} = \boldsymbol{d}^T\boldsymbol{E}\boldsymbol{c} \tag{1.70}$$

It should now be apparent that, if all the entries of a 2×2 matrix $\boldsymbol{A}$ are multiplied by the same scalar s, then its determinant is multiplied by s^2, i.e.,

$$\det(s\boldsymbol{A}) = s^2\det(\boldsymbol{A}) \tag{1.71}$$

The determinant of a 3×3 matrix $\boldsymbol{A}$ is defined below. To this end, we partition $\boldsymbol{A}$ column-wise as

$$\boldsymbol{A} = \begin{bmatrix}\boldsymbol{a}_1 & \boldsymbol{a}_2 & \boldsymbol{a}_3\end{bmatrix} \tag{1.72}$$

The definition of $\det(\boldsymbol{A})$ can be expressed as follows:

$$\det(\boldsymbol{A}) = \boldsymbol{a}_1 \times \boldsymbol{a}_2 \cdot \boldsymbol{a}_3 \tag{1.73}$$

The expression appearing in the right-hand side of the above equation is known as the *mixed product*, *triple product*, or even the *box product*, of the three given vectors, and is sometimes represented as

$$[\boldsymbol{a}_1,\ \boldsymbol{a}_2,\ \boldsymbol{a}_3] = \boldsymbol{a}_1 \times \boldsymbol{a}_2 \cdot \boldsymbol{a}_3 \tag{1.74}$$

Notice that this representation *resembles* that of a 3×3 matrix, as given in Eq. (1.72). The latter is common in *linear algebra*, and hence this is a matrix representation that we adopt here. The reader is alerted to the difference between the two representations: *the former carries no commas*.

The triple product admits a geometric interpretation: Shown in Fig. 1.18 are three vectors, $\boldsymbol{a}$, $\boldsymbol{b}$, and $\boldsymbol{c}$, whose triple product is the scalar $p = (\boldsymbol{a} \times \boldsymbol{b}) \cdot \boldsymbol{c}$. The cross product $\boldsymbol{a} \times \boldsymbol{b}$ is a vector normal to the plane of its factors, of magnitude $\|\boldsymbol{a} \times \boldsymbol{b}\|$

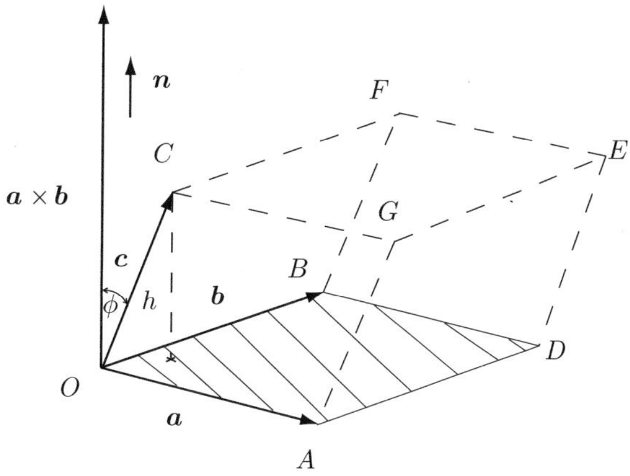

Fig. 1.18 The geometric interpretation of the box product

equal to the area A of the parallelogram defined by these factors and pointing in the direction that renders the triad $\{\boldsymbol{a},\ \boldsymbol{b},\ \boldsymbol{a} \times \boldsymbol{b}\}$ right-handed. We can thus write

$$p = A\boldsymbol{n} \cdot \boldsymbol{c} \tag{1.75}$$

where $\boldsymbol{n}$ is the unit vector derived from $\boldsymbol{a} \times \boldsymbol{b}$. Obviously, $|\boldsymbol{n} \cdot \boldsymbol{c}| = \|\boldsymbol{c}\||\cos\phi| = h$, the latter being the height of the parallelepiped $OADBCGEF$ defined by the three given vectors. Therefore, if V denotes the volume of the parallelepiped, then we have

$$|\boldsymbol{a} \times \boldsymbol{b} \cdot \boldsymbol{c}| = V \tag{1.76}$$

The volume of the tetrahedron $OABC$, in turn, is $V/6$.

Similar to relation (1.71), we have now, for a 3×3 matrix,

$$\det(s\boldsymbol{A}) = s^3 \det(\boldsymbol{A}) \tag{1.77}$$

Now we can define the determinant of a $n \times n$ matrix $\boldsymbol{A}$ for arbitrary n; this is done *recursively*, by defining this determinant as a *linear combination* of the determinants of $(n-1) \times (n-1)$ matrices. To this end, we denote the (i, j) entry of $\boldsymbol{A}$ as $a_{i,j}$. Moreover, the *minor* $M_{i,j}$ of entry $a_{i,j}$ is defined as the determinant of a $(n-1) \times (n-1)$ matrix $\boldsymbol{A}_{i,j}$, obtained from $\boldsymbol{A}$ upon deleting its ith row and its jth column, i.e.,

$$M_{i,j} = \det(\boldsymbol{A}_{i,j}) \tag{1.78}$$

Further, the *cofactor* $C_{i,j}$ of entry $a_{i,j}$ of $\boldsymbol{A}$ is nothing but $M_{i,j}$ itself if $i + j$ is even, $-M_{i,j}$ if $i + j$ is odd. That is,

$$C_{i,j} = (-1)^{i+j} M_{i,j} \tag{1.79}$$

Thus,

- The determinant of $\boldsymbol{A}$ is defined in terms of the ith row[6] as

$$\det(\boldsymbol{A}) = \sum_{j=1}^{n} a_{i,j} C_{i,j} \quad \forall \quad 1 \le i \le \mathrm{n} \tag{1.80}$$

- If we recall property (1.68), an alternative definition of $\det(\boldsymbol{A})$ follows, in terms of the jth column:

$$\det(\boldsymbol{A}) = \sum_{i=1}^{n} a_{i,j} C_{i,j} \quad \forall \quad 1 \le j \le \mathrm{n} \tag{1.81}$$

Actually, the definition of the determinant of a 3×3 matrix, Eq. (1.73), follows from the counterpart definition of a 2×2 determinant, Eq. (1.65), and the foregoing general definition of the determinant of a $n \times n$ matrix $\boldsymbol{A}$. Now we can state a generalization of relations (1.71) and (1.77):

The determinant of a $n \times n$ matrix $\boldsymbol{A}$ is homogeneous of degree n, i.e., if all the entries of $\boldsymbol{A}$ are multiplied by the same scalar s, then $\det(s\boldsymbol{A})$ *becomes*

$$\det(s\boldsymbol{A}) = s^n \det(\boldsymbol{A}) \tag{1.82}$$

As a result of relations (1.78) and (1.79), a property of the determinant follows[7]:

Property 1.4.1

If any two columns—rows—are interchanged, the absolute value of the determinant of an $n \times n$ matrix is preserved, but its sign is reversed.

As a consequence, we have

Property 1.4.2

The determinant of a 3×3 matrix is preserved under a cyclic permutation of its columns—rows.

As stated earlier, the computation of the determinant of a $n \times n$ matrix from its definition, Eq. (1.80) or, equivalently, Eq. (1.81), is extremely costly. Indeed, from the foregoing discussion it is apparent that computing a 2×2 determinant requires two multiplications and one addition, or, roughly, two *flops*. The computation of a 3×3 determinant requires the computation of three 2×2 *subdeterminants*, which amounts to, roughly, six flops, but then, each of these determinants (cofactors) must be multiplied by its corresponding entry, these three products being finally added up, which brings about three more flops—give or take one addition operation. Hence, the computation of a 3×3 determinant requires $3 \times 2 + 3 = 3! + 3 = 9$ flops. Using

[6] Any row, in fact.

[7] Its proof pertains to classical advanced books, e.g., Finkbeiner (1966).

induction to extrapolate a pattern, we can estimate that the computation of a $n \times n$ determinant consumes slightly over $n!$ flops. Now, the factorial grows extremely rapidly with n, which means that, even for moderately large values of n, $n!$ may lead to an unmanageably large number of flops.

As an example, let us consider a 30×30 matrix, which can easily arise in various engineering applications. The number N of flops required to compute the determinant of such a matrix would be, as obtained with computer algebra,

$$N = 30! = 265252859812191058636308480000000$$

which is a pretty large number. To gain insight into the size of this number, let us assume that we have an *OLCF-4* supercomputer,[8] capable of executing 200 pflops = 200×10^{15} flops, where pflops stands for *petaflops*, i.e., 2×10^{17} *floating-point operations per second*. Since the set of real numbers is a *continuum*—between any two given *distinct* real numbers, no matter how "close" to each other, there are infinitely many real numbers, no digital computer is capable of storing such numbers, never mind operating with them. A *floating-point number* is (a) computer-storable and (b) capable of being added to and multiplied by another *floating-point number*.

The time T such a supercomputer would take can now be readily found as

$$T = \frac{N}{2.00 \times 10^{17}} = 1.33 \times 10^{15} \text{ s} \tag{1.83}$$

which is, again, a pretty large time interval. In order to have an idea of how big this time estimate is, let us compare it with the age of the universe, which is about 1.38×10^{10} years.[9] In seconds, the age of the universe is 4.352×10^{17} s. Hence, the ratio r of T to the latest estimate of the age of the universe is 4.247×10^{3}. That is, the time the fastest computer would take to compute the determinant of a 30×30 matrix is, roughly, 0.4% the age of the universe.

The good news is that streamlined methods are available to compute determinants, when such a computation is at all needed. One method, studied in courses on numerical analysis and applied linear algebra, relies on what is known as the *LU-decomposition* of the $n \times n$ square matrix $\boldsymbol{A}$, under which this matrix is factored into the form $\boldsymbol{A} = \boldsymbol{L}\boldsymbol{U}$. In this factoring, $\boldsymbol{L}$ is a lower triangular matrix with only 1's on its diagonal and $\boldsymbol{U}$ is an upper triangular matrix. A property of the determinant states that *the determinant of a product of matrices equals the product of the determinants of the individual matrix factors*, and hence,

$$\det(\boldsymbol{A}) = \det(\boldsymbol{L})\det(\boldsymbol{U}) \tag{1.84}$$

By virtue of the structure of $\boldsymbol{L}$, we have $\det(\boldsymbol{L}) = 1$, and hence $\det(\boldsymbol{A}) = \det(\boldsymbol{U})$. Moreover, given that $\boldsymbol{U}$ is upper triangular, its determinant equals the product of

[8]The fastest supercomputer in the world, as per Lohr (2018).

[9]Castelvecchi (2019).

its diagonal entries, which consumes only $n-1$ multiplications. Additionally, the LU-decomposition of a $n \times n$ matrix requires M_n multiplications and A_n additions[10]:

$$M_n = \frac{n^3}{3} + \frac{n^2}{2} + \frac{n}{6}, \quad A_n = \frac{n^3}{3} - \frac{n}{3} \tag{1.85}$$

For a 30×30 matrix, the foregoing figures amount to 9 155 multiplications and 8 990 additions, or roughly 9 000 flops. Any modern PC can execute approximately 10^8 floating-point operations per second, which means that the computation of a 30×30 determinant consumes about 100 μs, quite a short time interval when compared to the age of the universe! In 100 μs light would travel only 30 km.

1.4.6.1 Determinants of Block Matrices

Computing determinants of $n \times n$ matrices, for $n > 3$, can be achieved by resorting to the formulas available for *block-partitioned matrices*. For example, in Chap. 4, we will need to find the determinant of a 4×4 matrix representing an *affine transformation* in three-dimensional space. In 3D, the affine transformation in question is given by a *homogeneous* 4×4 matrix, as introduced in Sect. 4.3. We thus consider here a $n \times n$ block matrix $\boldsymbol{P}$, where n is any natural number, defined by blocks, namely,

$$\boldsymbol{P} = \begin{bmatrix} \boldsymbol{A} & \boldsymbol{B} \\ \boldsymbol{C} & \boldsymbol{D} \end{bmatrix} \tag{1.86}$$

in which we assume that all blocks are *compatible*, i.e., if $\boldsymbol{A}$ is of $p \times p$ and $\boldsymbol{D}$ is of $q \times q$, then $\boldsymbol{B}$ is of $p \times q$ and $\boldsymbol{C}$ of $q \times p$. We thus have implicitly assumed that $p + q = n$.

As an example, consider the 4×4 *homogeneous-transformation matrix* $\boldsymbol{T}$ of Eq. (4.31), reproduced below for quick reference:

$$\boldsymbol{T} = \begin{bmatrix} \boldsymbol{M} & \boldsymbol{t} \\ \boldsymbol{0}^T & 1 \end{bmatrix} \tag{1.87}$$

In this case, the 3×3 matrix $\boldsymbol{M}$ represents a rotation, a reflection, or a scaling—these terms are explained in detail in Chap. 4—while the three-dimensional vector $\boldsymbol{t}$ represents a translation, $\boldsymbol{0}$ the three-dimensional zero vector and 1 is the real unity.

The formulas that allow the user to compute the determinant of the block matrix $\boldsymbol{P}$ given in Eq. (1.86) are displayed below[11]:

$$\det\left(\begin{bmatrix} \boldsymbol{A} & \boldsymbol{B} \\ \boldsymbol{C} & \boldsymbol{D} \end{bmatrix}\right) = \det(\boldsymbol{A})\det(\boldsymbol{D} - \boldsymbol{C}\boldsymbol{A}^{-1}\boldsymbol{B}) = \det(\boldsymbol{D})\det(\boldsymbol{A} - \boldsymbol{B}\boldsymbol{D}^{-1}\boldsymbol{C}) \tag{1.88}$$

---

[10]Dahlquist and Björck (1974).

[11]These formulas can be proven by various means; this proof not being pertinent to the book material, it is left aside.

Notice that any of the two foregoing formulas can be applied. However, the first formula requires that $\boldsymbol{A}$ be invertible, while the second that $\boldsymbol{D}$ be so. In some cases, one of these two matrices is invertible, but not both. The user must choose judiciously which of the two formulas to apply. If none of $\boldsymbol{A}$ and $\boldsymbol{D}$ is invertible, to compute $\det(\boldsymbol{P})$, a reshuffling of the blocks may be needed, while taking into account that the sign of a determinant is preserved only under a *cyclic permutation* of either its columns or its rows.

As an example, we obtain the determinant of the 4×4 matrix of Eq. (1.87), where we identify the blocks below:

$$\boldsymbol{A} = \boldsymbol{M}, \quad \boldsymbol{B} = \boldsymbol{t}, \quad \boldsymbol{C} = \boldsymbol{0}^T, \quad \boldsymbol{D} = 1 \tag{1.89a}$$

and hence, by application of the first of formulas (1.88), we have

$$\begin{aligned} \det(\boldsymbol{T}) &= \det(\boldsymbol{M})\det(1 - \boldsymbol{0}^T\boldsymbol{M}^{-1}\boldsymbol{t}) \\ &= \det(\boldsymbol{M})(1) = \det(\boldsymbol{M}) \end{aligned}$$

which shows that, regardless of the value of vector $\boldsymbol{t}$, the determinant of the 4×4 *homogeneous-transformation matrix* $\boldsymbol{T}$ is always identical to that of $\boldsymbol{M}$, a 3×3 matrix.

1.4.7 The Cross-Product Matrix

The cross product $\boldsymbol{a} \times \boldsymbol{b}$ appearing in Eq. (1.52) is reproduced below for quick reference:

$$\boldsymbol{a} \times \boldsymbol{b} = \begin{bmatrix} a_y b_z - a_z b_y \\ a_z b_x - a_x b_z \\ a_x b_y - a_y b_x \end{bmatrix} \tag{1.90}$$

which, as the reader is invited to verify, can be expressed as the product of a 3×3 matrix $\boldsymbol{A}$ times vector $\boldsymbol{b}$, namely,

$$\boldsymbol{a} \times \boldsymbol{b} = \underbrace{\begin{bmatrix} 0 & -a_z & a_y \\ a_z & 0 & -a_x \\ -a_y & a_x & 0 \end{bmatrix}}_{\boldsymbol{A}} \underbrace{\begin{bmatrix} b_x \\ b_y \\ b_z \end{bmatrix}}_{\boldsymbol{b}} \tag{1.91}$$

matrix $\boldsymbol{A}$ being termed the *cross-product matrix* of vector $\boldsymbol{a}$, represented as

$$\boldsymbol{A} = \mathrm{CPM}(\boldsymbol{a}) \tag{1.92}$$

It is left as an exercise to prove the relation

$$A^2 = -||\boldsymbol{a}||^2\mathbf{1} + \boldsymbol{a}\boldsymbol{a}^T \tag{1.93}$$

Note: The *external product* of two vectors, of the form $\boldsymbol{a}\boldsymbol{b}^T$, is formally introduced in Eq. (1.111).

Hence, vector $\boldsymbol{a} \times \boldsymbol{b}$ can be expressed as

$$\boldsymbol{a} \times \boldsymbol{b} = \boldsymbol{A}\boldsymbol{b} \tag{1.94}$$

1.4.8 The "Triple" Vector Product

Let $\boldsymbol{a}$, $\boldsymbol{b}$, and $\boldsymbol{c}$ be three vectors, a fourth vector $\boldsymbol{v}$, defined as

$$\boldsymbol{v} = (\boldsymbol{a} \times \boldsymbol{b}) \times \boldsymbol{c} \tag{1.95}$$

which is a *double vector product*. This vector is sometimes referred to in the literature as the "triple vector product," but this is apparently a misnomer, hence the quotation marks. While the proof of the formula is available in the literature,[12] our proof is based only on material covered in Sects. 1.1–1.3, which makes it fit in only 1.5 pp, as opposed to four in the cited reference.

In order to picture vector $\boldsymbol{v}$, we need some geometry: let Π_{ab} denote the plane defined by vectors $\boldsymbol{a}$ and $\boldsymbol{b}$, passing through the origin O; this plane is normal to $\boldsymbol{a} \times \boldsymbol{b}$, of course. Moreover, let Π_c denote the plane normal to $\boldsymbol{c}$.

Now, $\boldsymbol{v}$ is normal to $\boldsymbol{a} \times \boldsymbol{b}$, and hence lies in Π_{ab}. As $\boldsymbol{v}$ is normal to $\boldsymbol{c}$ as well, it also lies in Π_c. Therefore, $\boldsymbol{v}$ is parallel to line $\mathcal{L}$, the intersection of the two foregoing planes, as depicted in Fig. 1.19. As a consequence, $\boldsymbol{v}$ is a linear combination of $\boldsymbol{a}$ and $\boldsymbol{b}$, of the form

$$\boldsymbol{v} \equiv (\boldsymbol{a} \times \boldsymbol{b}) \times \boldsymbol{c} = k_a\boldsymbol{a} + k_b\boldsymbol{b} \tag{1.96}$$

As shown in Sect. 1.4.9, after some preliminary background has been introduced, the above coefficients are given as

$$k_a = -\boldsymbol{b} \cdot \boldsymbol{c}, \quad k_b = \boldsymbol{a} \cdot \boldsymbol{c} \tag{1.97}$$

We thus have the classic formula

$$(\boldsymbol{a} \times \boldsymbol{b}) \times \boldsymbol{c} = (\boldsymbol{a} \cdot \boldsymbol{c})\boldsymbol{b} - (\boldsymbol{b} \cdot \boldsymbol{c})\boldsymbol{a} \tag{1.98}$$

[12] Chapman and Milne (1939).

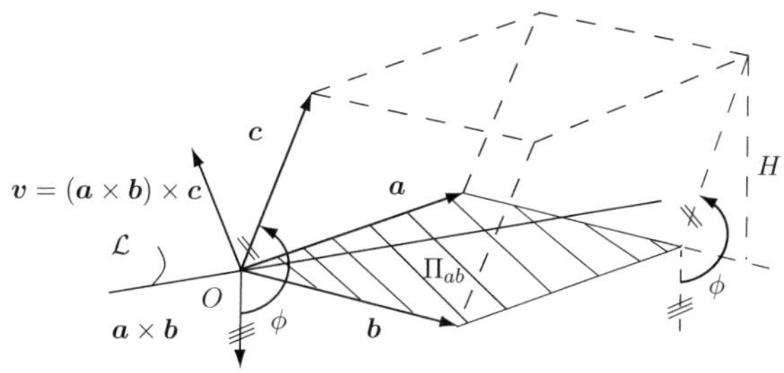

Fig. 1.19 Geometric interpretation of the double (a.k.a. "triple") vector product

1.4.9 Derivation of the "Double Vector-Product Formula"

Now we can prove formulas (1.97) upon finding expressions for its coefficients k_a and k_b. To this end, we impose first the condition that vector $\boldsymbol{v}$, the double cross product introduced in Eq. (1.96), be normal to $\boldsymbol{c}$, i.e.,

$$k_a(\boldsymbol{a}^T\boldsymbol{c}) + k_b(\boldsymbol{b}^T\boldsymbol{c}) = 0 \tag{1.99}$$

and hence

$$k_b = -\frac{\boldsymbol{a}^T\boldsymbol{c}}{\boldsymbol{b}^T\boldsymbol{c}}k_a \tag{1.100}$$

Therefore, $\boldsymbol{v}$ can now be expressed as

$$\boldsymbol{v} = k_a\frac{(\boldsymbol{b}^T\boldsymbol{c})\boldsymbol{a} - (\boldsymbol{a}^T\boldsymbol{c})\boldsymbol{b}}{\boldsymbol{b}^T\boldsymbol{c}} \tag{1.101}$$

Below we derive an expression for k_a in terms of $\boldsymbol{a}$, $\boldsymbol{b}$, and $\boldsymbol{c}$. To this end, the product $\boldsymbol{v}^T\boldsymbol{a}$ is computed *both* with the above expression for $\boldsymbol{v}$ *and* with the definition of $\boldsymbol{v}$, as given in Eq. (1.96):

$$\left[k_a\frac{(\boldsymbol{b}^T\boldsymbol{c})\boldsymbol{a} - (\boldsymbol{a}^T\boldsymbol{c})\boldsymbol{b}}{\boldsymbol{b}^T\boldsymbol{c}}\right]^T\boldsymbol{a} = [(\boldsymbol{a}\times\boldsymbol{b})\times\boldsymbol{c}]^T\boldsymbol{a} \tag{1.102}$$

Next, the right-hand side of Eq. (1.102) is first expanded upon exchanging the cross- and dot-product symbols:

$$[(\boldsymbol{a}\times\boldsymbol{b})\times\boldsymbol{c}]^T\boldsymbol{a} \equiv (\boldsymbol{a}\times\boldsymbol{b})^T(\boldsymbol{c}\times\boldsymbol{a}) \tag{1.103}$$

then, equated with the expansion of the left-hand side of Eq. (1.102), i.e.,

$$\frac{(\boldsymbol{b}^T\boldsymbol{c})(\boldsymbol{a}^T\boldsymbol{a}) - (\boldsymbol{a}^T\boldsymbol{c})(\boldsymbol{b}^T\boldsymbol{a})}{\boldsymbol{b}^T\boldsymbol{c}}k_a = (\boldsymbol{a}\times\boldsymbol{b})^T(\boldsymbol{c}\times\boldsymbol{a}) \tag{1.104}$$

which readily leads to

$$k_a = \frac{(\boldsymbol{b}^T\boldsymbol{c})(\boldsymbol{a}\times\boldsymbol{b})^T(\boldsymbol{c}\times\boldsymbol{a})}{(\boldsymbol{b}^T\boldsymbol{c})||\boldsymbol{a}||^2 - (\boldsymbol{a}^T\boldsymbol{c})(\boldsymbol{b}^T\boldsymbol{a})} \equiv \frac{N}{(\boldsymbol{b}^T\boldsymbol{c})||\boldsymbol{a}||^2 - (\boldsymbol{a}^T\boldsymbol{c})(\boldsymbol{b}^T\boldsymbol{a})} \tag{1.105}$$

Now, with $\boldsymbol{a}\times\boldsymbol{b}$ replaced with $\boldsymbol{Ab}$ and $\boldsymbol{c}\times\boldsymbol{a}$ with $-\boldsymbol{Ac}$, $\boldsymbol{A}\equiv \text{CPM}(\boldsymbol{a})$, N is expanded as

$$N = (\boldsymbol{b}^T\boldsymbol{c})(\boldsymbol{Ab})^T(-\boldsymbol{Ac}) \equiv (\boldsymbol{b}^T\boldsymbol{c})\boldsymbol{b}^T\boldsymbol{A}^T\boldsymbol{Ac} = (\boldsymbol{b}^T\boldsymbol{c})\boldsymbol{b}^T\boldsymbol{A}^2\boldsymbol{c} \tag{1.106}$$

Next, upon substitution of the identity (1.93) into the above expression for N, we obtain

$$N = (\boldsymbol{b}^T\boldsymbol{c})\left[-||\boldsymbol{a}||^2\boldsymbol{b}^T\boldsymbol{c} + (\boldsymbol{b}^T\boldsymbol{a})(\boldsymbol{a}^T\boldsymbol{c})\right] \tag{1.107}$$

thereby obtaining the expressions for k_a and k_b:

$$k_a = \frac{(\boldsymbol{b}^T\boldsymbol{c})\left[-||\boldsymbol{a}||^2\boldsymbol{b}^T\boldsymbol{c} + (\boldsymbol{b}^T\boldsymbol{a})(\boldsymbol{a}^T\boldsymbol{c})\right]}{(\boldsymbol{b}^T\boldsymbol{c})||\boldsymbol{a}||^2 - (\boldsymbol{a}^T\boldsymbol{c})(\boldsymbol{a}^T\boldsymbol{b})} \tag{1.108a}$$

$$k_b = -\frac{(\boldsymbol{a}^T\boldsymbol{c})\left[-||\boldsymbol{a}||^2\boldsymbol{b}^T\boldsymbol{c} + (\boldsymbol{b}^T\boldsymbol{a})(\boldsymbol{a}^T\boldsymbol{c})\right]}{(\boldsymbol{b}^T\boldsymbol{c})||\boldsymbol{a}||^2 - (\boldsymbol{a}^T\boldsymbol{c})(\boldsymbol{a}^T\boldsymbol{b})} \tag{1.108b}$$

Then, upon substitution of the above expressions into Eq. (1.96), $\boldsymbol{v}$ is obtained as

$$\boldsymbol{v} = \frac{(\boldsymbol{b}^T\boldsymbol{c})\left[-||\boldsymbol{a}||^2\boldsymbol{b}^T\boldsymbol{c} + (\boldsymbol{b}^T\boldsymbol{a})(\boldsymbol{a}^T\boldsymbol{c})\right]}{(\boldsymbol{b}^T\boldsymbol{c})||\boldsymbol{a}||^2 - (\boldsymbol{a}^T\boldsymbol{c})(\boldsymbol{a}^T\boldsymbol{b})}\frac{(\boldsymbol{b}^T\boldsymbol{c})\boldsymbol{a} - (\boldsymbol{a}^T\boldsymbol{c})\boldsymbol{b}}{\boldsymbol{b}^T\boldsymbol{c}} \tag{1.109}$$

which readily simplifies to

$$\boldsymbol{v} = (\boldsymbol{a}^T\boldsymbol{c})\boldsymbol{b} - (\boldsymbol{b}^T\boldsymbol{c})\boldsymbol{a} \tag{1.110}$$

thereby proving Eq. (1.97).

Further, given two three-dimensional vectors $\boldsymbol{a}$ and $\boldsymbol{b}$, their *external product* $\boldsymbol{ab}^T$ is a 3×3 matrix, defined as

$$\boldsymbol{ab}^T = \begin{bmatrix} a_1 \\ a_2 \\ a_3 \end{bmatrix}\begin{bmatrix} b_1 & b_2 & b_3 \end{bmatrix} = \begin{bmatrix} a_1b_1 & a_1b_2 & a_1b_3 \\ a_2b_1 & a_2b_2 & a_2b_3 \\ a_3b_1 & a_3b_2 & a_3b_3 \end{bmatrix} \tag{1.111}$$

With the foregoing definition, the "triple" vector product (1.98) can be expressed as

$$(\boldsymbol{a} \times \boldsymbol{b}) \times \boldsymbol{c} = \boldsymbol{M}\boldsymbol{c}, \quad \boldsymbol{M} \equiv \boldsymbol{b}\boldsymbol{a}^T - \boldsymbol{a}\boldsymbol{b}^T \tag{1.112}$$

1.4.10 Matrix Inversion

A $n \times n$ matrix whose determinant vanishes is termed *singular*; otherwise, the matrix is said to be *non-singular*. Non-singular matrices are sometimes referred to as *regular*.

Any $n \times n$ non-singular matrix $\boldsymbol{A}$ has an associated *inverse*, denoted $\boldsymbol{A}^{-1}$, such that

$$\boldsymbol{A}\boldsymbol{A}^{-1} = \boldsymbol{A}^{-1}\boldsymbol{A} = \mathbf{1} \tag{1.113}$$

where $\mathbf{1}$ denotes the $n \times n$ identity matrix.[13]

With the definition of cofactor introduced in Sect. 1.4.6, we can now define the *adjoint* Adj($\boldsymbol{A}$) of a $n \times n$ matrix $\boldsymbol{A}$ as the $n \times n$ matrix whose (i, j) entry is the cofactor $C_{i,j}$ of $a_{i,j}$, namely,

$$[\text{Adj } (\boldsymbol{A})]_{i,j} = C_{i,j} \tag{1.114}$$

Now, the inverse of $\boldsymbol{A}$ can be computed using the formula

$$\boldsymbol{A}^{-1} = \frac{1}{\det(\boldsymbol{A})}\text{Adj}(\boldsymbol{A}) \tag{1.115}$$

In reality, the matrix inverse is seldom needed, as such, to perform computations in practical engineering problems, but it occurs frequently in analysis. Indeed, the matrix inverse occurs when solving a system of n linear equations in n unknowns. In these cases, the numerical procedure relies on the LU-decomposition of the matrix coefficient and the observation that a *triangular* system of equations admits a *recursive* solution involving only arithmetic operations—additions and subtractions. A system of equations is considered to be *upper triangular* if the nth equation involves only the nth unknown, the $(n-1)$st equation only the nth and the $(n-1)$st unknowns, and so on, with the first equation involving all unknowns:

$$\begin{aligned} a_{11}x_1 + a_{12}x_2 + a_{13}x_3 + \cdots + a_{1n}x_n &= b_1 \\ a_{22}x_2 + a_{23}x_3 + \cdots + a_{2n}x_n &= b_2 \\ \ddots \quad \vdots &= \vdots \\ a_{mn}x_n &= b_n \end{aligned} \tag{1.116}$$

[13]Such a matrix is, sometimes, referred to as a "unit" or, even, as a "unitary" matrix, but this is misleading, as a "unitary matrix" has a precise definition in linear algebra, namely, as the counterpart of an *orthogonal matrix* when the matrix is defined over the *complex field*, i.e., the set of the complex numbers. This is not the case in the book.

A *lower triangular* system is defined likewise. Therefore, an upper triangular system of linear equations can be readily solved *recursively* by *backward substitution*: Start by solving the nth equation for the nth unknown, thereby ending up with only $n-1$ unknowns left. The $(n-1)$st equation is next solved for the $(n-1)$st unknown, which leaves us with only $n-2$ unknowns to compute. At the beginning of the nth *recursion*, we are left with only one unknown, x_1, which can readily be solved for from the first equation, the only equation left.

We will not elaborate further on the solution of linear systems of equations for arbitrary values of n, but will rather focus on two special cases that can be handled *symbolically*, i.e., without a numerical procedure, using formulas instead. Obviously, the simplest non-trivial cases occur when $n = 2$ and $n = 3$, all we need in this book, as discussed below.

As the reader can readily verify, for a 2×2 matrix $\boldsymbol{A}$, partitioned as shown in Eq. (1.64),

$$\boldsymbol{A}^{-1} = \frac{1}{\det(\boldsymbol{A})}\begin{bmatrix} \boldsymbol{b}^T \\ -\boldsymbol{a}^T \end{bmatrix}\boldsymbol{E} = \frac{1}{\det(\boldsymbol{A})}\boldsymbol{E}\begin{bmatrix} -\boldsymbol{d} & \boldsymbol{c} \end{bmatrix} \tag{1.117}$$

where $\boldsymbol{E}$ was defined in Eq. (1.54).

A quick verification involves only the computation of the product $\boldsymbol{A}\boldsymbol{A}^{-1}$, or $\boldsymbol{A}^{-1}\boldsymbol{A}$ for that matter, which should yield the 2×2 identity matrix.

Given a 3×3 matrix $\boldsymbol{A}$ partitioned as in Eq. (1.72), its inverse can be evaluated in the form:

$$\boldsymbol{A}^{-1} = \frac{1}{\Delta}\begin{bmatrix} (\boldsymbol{a}_2 \times \boldsymbol{a}_3)^T \\ (\boldsymbol{a}_3 \times \boldsymbol{a}_1)^T \\ (\boldsymbol{a}_1 \times \boldsymbol{a}_2)^T \end{bmatrix}, \quad \Delta \equiv \det(\boldsymbol{A}) = \boldsymbol{a}_1 \times \boldsymbol{a}_2 \cdot \boldsymbol{a}_3 \tag{1.118}$$

Again, the reader can verify the validity of the foregoing formula by straightforward computation of the product $\boldsymbol{A}\boldsymbol{A}^{-1}$ or, equivalently, of $\boldsymbol{A}^{-1}\boldsymbol{A}$.

1.4.10.1 Inverses of Block Matrices

Given the same block matrix as in Eq. (1.86), its inverse is given by

$$\boldsymbol{P}^{-1} = \begin{bmatrix} \boldsymbol{X} & \boldsymbol{Y} \\ \boldsymbol{Z} & \boldsymbol{U} \end{bmatrix} \tag{1.119}$$

with $\boldsymbol{X}, \boldsymbol{Y}, \boldsymbol{Z}$, and $\boldsymbol{U}$ being, correspondingly, $p \times p$, $p \times q$, $q \times p$, and $q \times q$ blocks, whose values are given below:

$$\boldsymbol{X} = (\boldsymbol{A} - \boldsymbol{B}\boldsymbol{D}^{-1}\boldsymbol{C})^{-1} \tag{1.120a}$$

$$\boldsymbol{U} = (\boldsymbol{D} - \boldsymbol{C}\boldsymbol{A}^{-1}\boldsymbol{B})^{-1} \tag{1.120b}$$

$$\boldsymbol{Y} = -\boldsymbol{A}^{-1}\boldsymbol{B}\boldsymbol{U} \tag{1.120c}$$

$$\boldsymbol{Z} = -\boldsymbol{D}^{-1}\boldsymbol{C}\boldsymbol{X} \tag{1.120d}$$

Sure enough, the integers p and q must obey

$$p + q = n$$

The validity of the foregoing formulas can be verified by straightforward computations: simply multiply matrix $\boldsymbol{P}$, as given in Eq. (1.86), by $\boldsymbol{P}^{-1}$, as given in Eq. (1.119). The product should yield the $n \times n$ identity matrix.

As an exercise, let us compute the inverse of the 4×4 homogeneous-transformation matrix $\boldsymbol{T}$ of Eq. (1.87). In this case, we have, for $\boldsymbol{X}$ and $\boldsymbol{U}$,

$$\boldsymbol{X} = [\boldsymbol{M} - \boldsymbol{t}(1)^{-1}\boldsymbol{0}^T]^{-1}$$
$$\boldsymbol{U} = (1 - \boldsymbol{0}^T\boldsymbol{M}^{-1}\boldsymbol{t})^{-1}$$

In the above expressions, notice that the $\boldsymbol{D}$ block in matrix $\boldsymbol{T}$ is the real unity 1, which can be interpreted as the 1×1 "identity matrix," its inverse being the real unity itself. Moreover, $\boldsymbol{C}$ is $\boldsymbol{0}^T$, the transpose of the three-dimensional zero vector. Hence, the product $\boldsymbol{B}\boldsymbol{D}^{-1}\boldsymbol{C}$, i.e., $\boldsymbol{t}(1)^{-1}\boldsymbol{0}^T$ in the brackets of the expression for $\boldsymbol{X}$, becomes

$$\boldsymbol{t}(1)^{-1}\boldsymbol{0}^T \equiv \boldsymbol{t}\boldsymbol{0}^T$$

which, as the reader can verify, is the 3×3 zero matrix $\boldsymbol{O}$, and hence

$$\boldsymbol{X} = \boldsymbol{M}^{-1},$$

i.e., the inverse of $\boldsymbol{M}$. The latter exists because, as per the description given right below Eq. (1.87), $\boldsymbol{M}$ represents a rotation, a reflexion or a scaling, which are necessarily invertible. By the same token, the reader can verify that the product $\boldsymbol{0}^T\boldsymbol{M}^{-1}\boldsymbol{t}$ in the parenthesis of the expression for $\boldsymbol{U}$ reduces to the 1×1 "zero matrix," i.e., the real 0. As a consequence,

$$\boldsymbol{U} = \boldsymbol{1}^{-1} = \boldsymbol{1}$$

Therefore,

$$\boldsymbol{Y} = -\boldsymbol{M}^{-1}\boldsymbol{t}\boldsymbol{U} = -\boldsymbol{M}^{-1}\boldsymbol{t}$$

and

$$\boldsymbol{Z} = -\boldsymbol{D}^{-1}\boldsymbol{C}\boldsymbol{X} = -1^{-1}\boldsymbol{0}^T\boldsymbol{M}^{-1} = \boldsymbol{0}^T$$

Finally, substituting all four expressions for $\boldsymbol{X}$, $\boldsymbol{Y}$, $\boldsymbol{Z}$, and $\boldsymbol{U}$ into Eq. (1.119), we obtain the desired inverse:

$$\boldsymbol{T}^{-1} = \begin{bmatrix} \boldsymbol{M}^{-1} & -\boldsymbol{M}^{-1}\boldsymbol{t} \\ \boldsymbol{0}^T & 1 \end{bmatrix} \tag{1.121}$$

That is, $\boldsymbol{T}$ and its inverse bear the same *gestalt*: the two lower blocks do not change, while the left-upper block becomes the inverse of the corresponding block in $\boldsymbol{T}$, the right-upper block becoming the negative of the product of the left-upper block in $\boldsymbol{T}^{-1}$ by the right-upper block in $\boldsymbol{T}$.

In summary, computing the inverse of a 4×4 homogeneous-transformation matrix, and of any 4×4 matrix for that matter, reduces to computing the inverse of a 3×3 matrix when the formulas (1.120a)–(1.120d) are invoked. Since we have a formula for the inverse of a 3×3 matrix in Eq. (1.115), it is straightforward to obtain a formula for the inverse of any particular 4×4 matrix. Finally, notice that the same formulas can be applied to compute the inverse of the 3×3 homogeneous-transformation matrix $\boldsymbol{T}$ introduced in Sect. 4.1.

1.5 Summary

The geometric and linear-algebraic tools essential for the construction of geometric objects were provided in this chapter. A connection between the two essential components of graphic communication in design tasks, regardless of the context—industrial design, architecture, civil engineering, mechanical engineering, etc.—is included. These components pertain to *art* and *technology*, activities that cannot be separated in any design task. Art is needed at the *preliminary design* stage, in which the designer starts with the specific need of the client. The first task faced by the designer is to produce an *embodiment* of the object to be proposed to the client to satisfy her or his need. The importance of free-hand sketching at this stage cannot be overstated. The designer's task here is to produce some alternative preliminary, and attractive, solutions allowing the client to visualize the object to be expected at the end of the design work, i.e., the *deliverable* in the art terminology.

It is then up to the client and the designer to agree on the embodiment that best satisfies the client's need. In the next stage of the design work, the *detailed design* of this variant is to be conducted. At this stage, drawings are needed for the production of the designed object—a civil engineering structure, a machine or a consumption good. These are *geometrically sound* drawings, to be produced using a modern CAD (computer-aided design) software system. While modern CAD systems offer virtually unlimited possibilities of creating sound geometric forms, the CAD-system user—a technician, if not the designer herself—must be highly conversant in the mathematical tools available in the system. The chapter includes an introduction to the fundamental concepts of linear algebra, such as vectors and matrices, and their manipulation, to solve the systems of equations most frequently encountered in geometry construction. An invaluable concept in the manipulation of vector and matrix arrays is the organization of the arrays in blocks. An introduction to the algebraic manipulation of arrays of blocks is included, to enable the reader to solve "large" systems of equations[14] that would require specialized knowledge, such as the LU-decomposition of square matrices, beyond the scope of the book. The level of difficulty of the exercises at the end of the chapter is commensurate with the material included in the chapter. The reader may appreciate that, without the tools provided in this chapter, some of the exercises included would be insurmountably difficult to handle.

1.6 Exercises

1.1 Consider the three-dimensional vector $\boldsymbol{p}_1 = [0 \quad 3 \quad 2]^T$.

(a) Find one of the two unit vectors $\boldsymbol{p}_2$ normal to $\boldsymbol{p}_1$ and to the X-axis.
(b) Find a unit vector normal to the plane $\mathcal{P}$ containing $\boldsymbol{p}_1$ and $\boldsymbol{p}_2$.

1.2 (a) Compute the determinant of the block matrix

$$\boldsymbol{M}_1 = \begin{bmatrix} \boldsymbol{A} & \boldsymbol{B} \\ \boldsymbol{C} & \boldsymbol{D} \end{bmatrix},$$

where
$\boldsymbol{A} = \begin{bmatrix} 1 & 4 \\ 2 & 3 \end{bmatrix}$, $\boldsymbol{B} = \begin{bmatrix} 3 \\ 1 \end{bmatrix}$, $\boldsymbol{C} = \begin{bmatrix} 4 & 1 \end{bmatrix}$, and $\boldsymbol{D} = [2]$.

(b) Compute the determinant of the matrix

$$\boldsymbol{M}_2 = \begin{bmatrix} 2 & 8 & 6 \\ 4 & 6 & 2 \\ 8 & 2 & 4 \end{bmatrix}.$$

1.3 (a) Compute the inverse of matrix $\boldsymbol{M}_1$ of Exercise 1.2(a).
(b) Compute the inverse of matrix $\boldsymbol{M}_2$ of Exercise 1.2(b).

1.4 Given the triplet $\{\boldsymbol{v}_1, \boldsymbol{v}_2, \boldsymbol{v}_3\}$:

$$\boldsymbol{v}_1 = \begin{bmatrix} 1 \\ 0 \\ 0 \end{bmatrix}, \quad \boldsymbol{v}_2 = \begin{bmatrix} 1 \\ 1 \\ 0 \end{bmatrix}, \quad \boldsymbol{v}_3 = \begin{bmatrix} -1 \\ -1 \\ 1 \end{bmatrix}$$

(a) Is the triplet coplanar?
(b) If the answer to the above question is **No**, define the matrix $\boldsymbol{V} = [\boldsymbol{v}_1 \quad \boldsymbol{v}_2 \quad \boldsymbol{v}_3]$ and compute $\boldsymbol{V}^{-1}$.

[14] The quoted qualifier is intended to stress that the "size" of a system of equations is relative, depending on the tools at hand. The book catering to beginners, a large system of linear equations would be one involving 10 or less equations in as many unknowns.

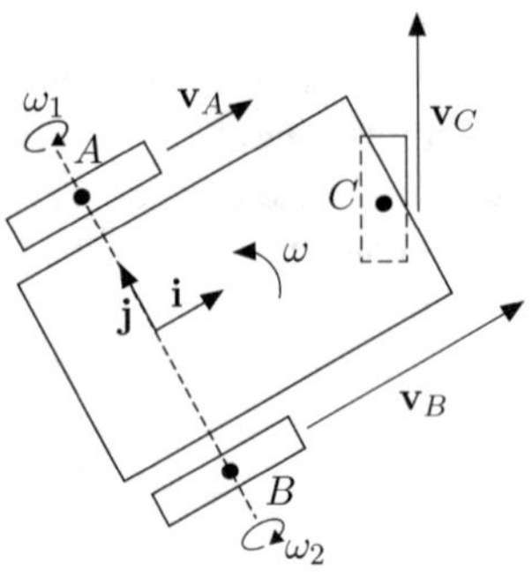

Fig. 1.20 Top view of a wheeled mobile robot

1.5 Three points are given below:

$$P_1(2, 0, 0); \quad P_2(0, 1, 0); \quad P_3(0, 0, 3)$$

(a) Find the unit normal $\mathbf{n}$ to the plane defined by the three points, with $\mathbf{n}$ pointing in the direction given by the *right-hand rule* when the points are visited in the order $P_1 P_2 P_3$.
(b) Find $\angle P_1 P_2 P_3$, which is measured positive in the direction of $\mathbf{n}$.

1.6 The top view of a mobile robot is shown in Fig. 1.20; the robot is driven by two wheels actuated by independent motors at angular speeds ω_1 and ω_2, respectively. A third wheel, carried by a bracket pinned at a vertical axis $\mathcal{A}$ passing through point C, is of the *caster* type. This wheel is supported by a bracket that allows it to turn freely about a vertical axis that passes through C and about its own axis, which is offset from the first axis. For purposes of simplicity, the offset is not shown in the figure.
Under the assumption that all wheels roll without sliding, the centers A and B move with velocities $\boldsymbol{v}_A$ and $\boldsymbol{v}_B$ given by

$$\boldsymbol{v}_A = \omega_1 r \boldsymbol{i}, \quad \boldsymbol{v}_B = \omega_2 r \boldsymbol{i}$$

where r is the radius of all three wheels. If the wheels are further assumed that they make contact with ground at points A and B a distance ℓ apart, then $\boldsymbol{v}_A$ and $\boldsymbol{v}_B$ obey the relation

$$\boldsymbol{v}_B = \boldsymbol{v}_A + \omega \boldsymbol{E}(\boldsymbol{b} - \boldsymbol{a}), \quad \boldsymbol{b} - \boldsymbol{a} = -\ell \boldsymbol{j}$$

with matrix $\boldsymbol{E}$ introduced in Sect. 1.4.5. Find an expression for ω in terms of ω_1, ω_2, r and ℓ. *Hint: recall that* $\mathbf{E}$ *transforms vectors in the plane into their perpendicular counterparts.*

1.7 A certain unit vector $\boldsymbol{v} = [v_1 \quad v_2 \quad v_3]^T$ is known to make angles of 60° and 120° with vectors $\boldsymbol{a}_1 = [1 \quad 0 \quad 0]^T$ and $\boldsymbol{a}_2 = [1 \quad 2 \quad 0]^T$, respectively. Find as many vectors $\boldsymbol{v}$ as possible that verify the above conditions and give a geometric interpretation of the exercise.

1.8 Given the three vectors

$$a_1 = \begin{bmatrix} 1 \\ 2 \\ 3 \end{bmatrix}, \quad a_2 = \begin{bmatrix} 4 \\ 5 \\ 6 \end{bmatrix}, \quad \text{and} \quad a_3 = \begin{bmatrix} 7 \\ 8 \\ 9 \end{bmatrix},$$

find an orthonormal triad $\{e_1, e_2, e_3\}$, i.e., a triad of mutually orthogonal unit vectors, such that (i) e_1 is parallel to and pointing in the same direction as a_1; (ii) e_2 lies in the plane of a_1 and a_2; and (iii) the triad is right-handed—if axes X, Y, and Z are defined parallel to and pointing in the same direction as e_1, e_2, and e_3, respectively, then the axes form a right-handed coordinate frame.

1.9 Solve the equation

$$a \times v + v = b$$

for v, with a and b given by

$$a = \begin{bmatrix} 1 \\ 2 \\ 3 \end{bmatrix}, \quad b = \begin{bmatrix} 3 \\ 2 \\ 1 \end{bmatrix}.$$

1.10 Computing determinants and inverses of *large* square matrices is sometimes possible, e.g., when the matrix is triangular. For M given below, compute (i) $\det(M)$ and (ii) M^{-1}.

$$M \equiv \begin{bmatrix} 1 & 2 & 3 & 4 & 5 \\ 0 & 2 & 3 & 4 & 5 \\ 0 & 0 & 3 & 4 & 5 \\ 0 & 0 & 0 & 4 & 5 \\ 0 & 0 & 0 & 0 & 5 \end{bmatrix}$$

1.11 Another example of a *large* matrix whose determinant and inverse are simple to compute is given below:

$$M \equiv \begin{bmatrix} \mathbf{1} & O \\ C & \mathbf{1} \end{bmatrix}, \quad \text{where} \quad C \equiv \begin{bmatrix} 0 & -1 & 1 \\ 1 & 0 & -1 \\ -1 & 1 & 0 \end{bmatrix}.$$

and $\mathbf{1}$ and O denote the 3×3 identity and zero matrices, respectively.

(i) Find $\det(M)$.
(ii) Find M^{-1}.

1.12 It is claimed that the three points given below are collinear:

$$P_1(1, 2, 3); \quad P_2(4, 5, 6); \quad P_3(3, 4, 5)$$

Prove or disprove the claim.

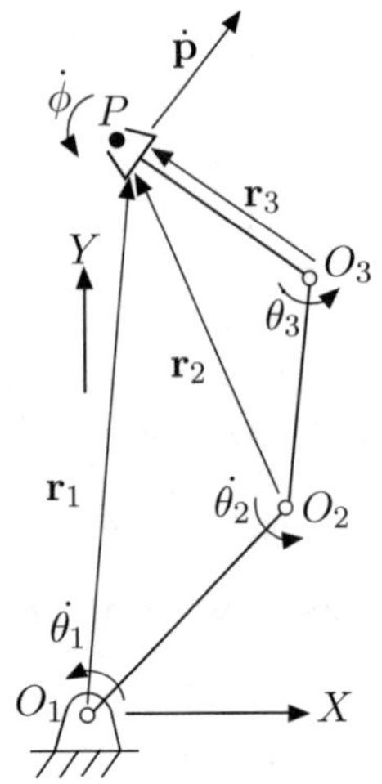

Fig. 1.21 A planar, three-axis robot

1.13 Shown in Fig. 1.21 is a planar, three-axis robot. The robot is driven by motors mounted on the joint axes O_1, O_2, and O_3, which produce computer-controlled rates $\dot{\theta}_1$, $\dot{\theta}_2$, and $\dot{\theta}_3$, respectively. These are mapped by the Jacobian matrix $\boldsymbol{J}$ into the three-dimensional end-effector *twist* $\boldsymbol{t}$. Vectors $\dot{\boldsymbol{\theta}}$ and $\boldsymbol{t}$ are defined as

$$\dot{\boldsymbol{\theta}} = \begin{bmatrix} \dot{\theta}_1 \\ \dot{\theta}_2 \\ \dot{\theta}_3 \end{bmatrix}, \quad \boldsymbol{t} = \begin{bmatrix} \dot{\phi} \\ \dot{\boldsymbol{p}} \end{bmatrix}$$

where $\boldsymbol{p} = \begin{bmatrix} x & y \end{bmatrix}^T$ is the position vector of the *operation point* P, while ϕ is the angle that the segment $\overline{O_3 P}$ of the end-effector makes with the X-axis. The foregoing vectors are related by

$$\boldsymbol{J}\dot{\boldsymbol{\theta}} = \boldsymbol{t}$$

while $\boldsymbol{J}$ is given by

$$\boldsymbol{J} = \begin{bmatrix} 1 & 1 & 1 \\ \boldsymbol{E}\boldsymbol{r}_1 & \boldsymbol{E}\boldsymbol{r}_2 & \boldsymbol{E}\boldsymbol{r}_3 \end{bmatrix}$$

and $\boldsymbol{E}$ is the 2×2 matrix introduced in Sect. 1.4.5. The computer control of the robot requires the inversion of matrix $\boldsymbol{J}$, and hence the robot becomes "handicapped" whenever $\boldsymbol{J}$ becomes singular. To detect when this occurs, $\det(\boldsymbol{J})$ is needed. **Without resorting to components**, find an expression for $\det(\boldsymbol{J})$ in terms of $\boldsymbol{E}$ and $\boldsymbol{r}_i$, for $i = 1, 2, 3$. *Hint: expand* $\det(\boldsymbol{J})$ *in terms of the cofactors of its first row and use relations* (1.67).

References

Castelvecchi D (2019) Into the dark ages. Nature 572:298–301

Chapman S, Milne EA (1939) The proof of the formula for the vector triple product. Math Gaz 23(253):35–38

Dahlquist G, Björck Å (1974) Numerical methods. Prentice-Hall Inc., Englewood Cliffs

Finkbeiner DT II (1966) Introduction to matrices and linear transformations. W.H. Freeman and Company, San Francisco

Lohr S (2018) Move over China, US is again home to world's speediest supercomputer. New York Times, 19 July

Chapter 2
2D Objects

Geometric elements are categorized as points, lines, surfaces, and solids. Surfaces and solids also have many subcategories. Points, lines, circles, and arbitrary curves are the basic 2D geometric primitives, or *generators*, from which other, more complex geometric objects can be derived or algorithmically produced. For example, by taking a straight line and moving it through a circular path lying in a plane normal to the line, while keeping the line normal to the plane, one can create a cylinder. This chapter defines, illustrates, and describes how to create points, lines, circles, polygons, polygonals, and arbitrary curves in the plane. The concept of *rigid body* in the plane and its properties is given due attention.

2.1 Points

A *point* is the simplest of the elementary geometric objects. Points are the basic building blocks for all other geometric objects. Points are indispensable when we create computer graphic displays and geometric models.

A point is a geometric concept that has position but no dimensions. A point position is defined by a set of real numbers, which are commonly referred to as *coordinates*. In the XY-plane, a point is represented by a pair of numbers, its *Cartesian coordinates* $(x,\ y)$, where x and y are the signed distances from the Y- and the X-axes, respectively.

The location of point P may also be expressed as an array of numbers, known as the *position vector* $\boldsymbol{p}$ of P, namely,

$$\boldsymbol{p} = \begin{bmatrix} x \\ y \end{bmatrix} \tag{2.1}$$

J. Angeles and D. Pasini, *Fundamentals of Geometry Construction*, Springer Tracts in Mechanical Engineering, https://doi.org/10.1007/978-3-030-43131-0_2

Note As stated previously, all vectors in this book are assumed to be *column arrays*. Moreover, *row arrays* are obtained from column arrays by *transposition*, which is indicated by a right superscript T, as defined in Chap. 1. We sometimes need to convert a column array to a row array for certain calculations.

2.2 Lines

Definition

A *line* in geometry and a line in CAD software are distinct concepts. In geometry, a line is a primitive that has a *location* and a *direction*, but no thickness. A line in geometry can be thought of as being generated by a point moving in a constant direction. Moreover, in geometry, a line is *unbounded*, in that it has no beginning and no end.

In CAD software, a line is, in fact, a segment, as defined by two *end points*.

Two elements of 2D geometry can define a line in the plane:

- two points.
- one point and one unit vector that indicates the direction of the line. The unit vector in the plane carries two *components*, say λ and μ, which are real numbers obeying the condition
$$\lambda^2 + \mu^2 = 1 \tag{2.2}$$
thereby imposing one *constraint* on the two numbers.
- one point and one unit vector perpendicular to the line. Again, the two components of the unit vector are related by condition (2.2).

Algebraic Representations of the Line

Explicit Representation:

$$y = mx + p \tag{2.3}$$

This is the well-known slope-intercept form, where m and p are the slope and the Y-axis intercept, i.e., the intersection point of the line with the Y-axis.

Implicit Representation:

$$Ax + By + C = 0, \quad A^2 + B^2 > 0 \tag{2.4}$$

where A, B, and C are constants. The explicit representation is expressed in the implicit form by substituting: $m = -A/B$ and $p = -C/B$, but this transformation requires $B \neq 0$, which is not always the case. Hence, the implicit representation is more general than its explicit counterpart.

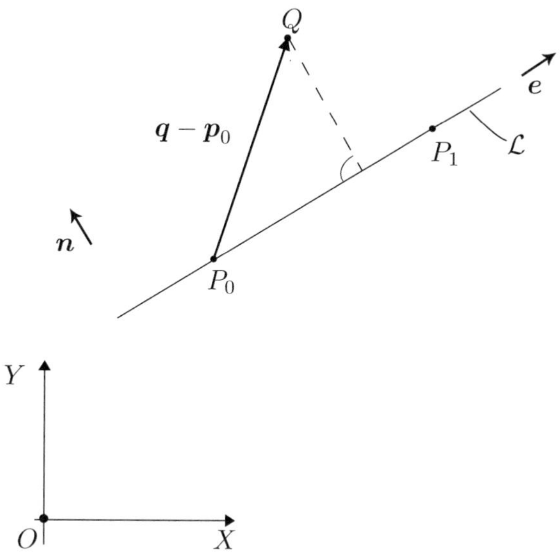

Fig. 2.1 Distance from a point Q to a line $\mathcal{L}$, both lying in the X–Y-plane

Parametric Representation:

$$x = au + b, \quad y = cu + d$$

where u is the *parameter*, and a, b, c, d are constants. Moreover, u is unbounded, i.e., $-\infty < u < \infty$.

2.2.1 Distance from a Point to a Line

Consider the line $\mathcal{L}$ given by Eq. (2.4) and depicted in Fig. 2.1. We want to compute the distance from a *given* point $Q(\xi, \eta)$ to the line. To this end, let us locate an arbitrary point $P_0(x_0, y_0)$ on the line. Since P_0 lies on the line, we cannot arbitrarily assign values to its coordinates. These must obey Eq. (2.4):

$$Ax_0 + By_0 + C = 0 \tag{2.5}$$

from which we can solve for either y_0 in terms of x_0 or the other way around. *To reduce roundoff errors, it is advisable to solve for the unknown multiplied by the coefficient with the higher absolute value.* Once we have one unknown in terms of the other, all we need to do is assign an arbitrary value to the latter, which will thus produce a pair (x_0, y_0) that complies with Eq. (2.5). Upon assigning a second numerical value to the same unknown, we should be able to produce a second pair (x_1, y_1) that also verifies Eq. (2.5). We can now define two points P_0 and P_1 in the plane, of position vectors $\boldsymbol{p}_0$ and $\boldsymbol{p}_1$, given by

$$\boldsymbol{p}_0 = \begin{bmatrix} x_0 \\ y_0 \end{bmatrix}, \quad \boldsymbol{p}_1 = \begin{bmatrix} x_1 \\ y_1 \end{bmatrix} \tag{2.6}$$

Next, we produce a *unit vector* $\boldsymbol{e}$ parallel to the line:

$$\boldsymbol{e} = \frac{\boldsymbol{p}_1 - \boldsymbol{p}_0}{\|\boldsymbol{p}_1 - \boldsymbol{p}_0\|} \tag{2.7}$$

Now, the unit normal $\boldsymbol{n}$ to the line can be most readily obtained by means of the $\boldsymbol{E}$ matrix introduced in Eq. (1.54):

$$\boldsymbol{n} = \boldsymbol{E}\boldsymbol{e} \tag{2.8}$$

As the reader can readily verify, the distance d sought is simply

$$d = |\boldsymbol{n}^T(\boldsymbol{q} - \boldsymbol{p}_0)| \tag{2.9}$$

where $\boldsymbol{q}$ is the position vector of $Q(\xi, \eta)$.

An alternative representation of the line can be obtained in terms of the normal to the line. From Fig. 2.1, it is apparent that

$$\boldsymbol{n}^T(\boldsymbol{p} - \boldsymbol{p}_0) = 0 \tag{2.10}$$

where point P_1 of the abovementioned figure has been replaced by a generic point P, of position vector $\boldsymbol{p}$. Moreover, let

$$\boldsymbol{n} = \begin{bmatrix} \lambda \\ \mu \end{bmatrix}, \quad \lambda^2 + \mu^2 = 1$$

and hence Eq. (2.10) leads to

$$\lambda(x - x_0) + \mu(y - y_0) = 0$$

or, upon expansion,

$$\lambda x + \mu y + \lambda x_0 - \mu y_0 = 0 \tag{2.11}$$

If now both sides of Eq. (2.4) are divided by $\sqrt{A^2 + B^2}$, one obtains

$$\frac{A}{\sqrt{A^2 + B^2}}x + \frac{B}{\sqrt{A^2 + B^2}}y + \frac{C}{\sqrt{A^2 + B^2}} = 0 \tag{2.12}$$

Upon comparing the left-hand sides of Eqs. (2.11) and (2.12), one finds

$$\lambda = \frac{A}{\sqrt{A^2+B^2}}, \quad \mu = \frac{B}{\sqrt{A^2+B^2}}, \quad \lambda x_0 + \mu y_0 = \frac{C}{\sqrt{A^2+B^2}} \tag{2.13}$$

a geometric interpretation then following for coefficients A, B, and C.

2.3 Planar Geometry and Polygons

A polygon is a multi-sided planar figure of any number of sides. In general, the polygon need not be a closed figure; however, unless otherwise stated, we will assume that the polygon in question is closed.[1] If the sides of the polygon are equal in length and all its internal angles are equal, the polygon is known as a *regular polygon*.

A polygon with n edges is given by an ordered set of points $P_1, P_2, \ldots, P_n$; the said polygon has *edge vectors* $\boldsymbol{v}_i = \boldsymbol{p}_{i+1} - \boldsymbol{p}_i$, for $i = 1, \ldots, n-1$, which connect pairs of neighboring points to form the desired polygon, with $\boldsymbol{p}_i$ denoting the position vector of P_i.

Note The number of vertices equals the number of edges.

One more interesting property: A regular n-sided *convex* polygon has a sum of interior angles I equal to

$$I = (n-2)\pi \tag{2.14}$$

Now we introduce a useful concept, *convexity*, which is quite general, but we restrict the concept in this chapter to planar figures bounded by closed contours.

Definition 2.3.1 *(Convexity)* Consider two arbitrary points P_1 and P_2 of an arbitrary planar closed figure, whose boundary is arbitrary as well, i.e., it can have line or arbitrary curved segments. If all the points of the segment $\overline{P_1P_2}$ lie either inside the figure or on its bounding contour, then the figure is said to be *convex*. A figure that does not have this property is called *non-convex*.

A general classification involves convex and non-convex polygons. Moreover, there exist many different types of polygons, regular polygons being defined as those that are

equilateral, which means that all sides are of equal length and
equiangular, which means that all interior angles at the vertices are equal.

The polygons that exhibit these characteristics are also referred to as *n-gons*, where n indicates the number of edges. Thus, an equilateral triangle is a 3-gon (or trigon), a square is a 4-gon (or tetragon), and so on.

---

[1] Sometimes, an open polygon is referred to as a *polygonal*.

Regular polygons are distinguished among themselves by their number of sides. We thus have, besides the familiar equilateral triangle and the square, the pentagon (five sides), the hexagon (six sides), the heptagon (seven sides), the octagon (eight sides), the nonagon (nine sides), the decagon (10 sides), the hendecagon or undecagon (11 sides), the dodecagon (12 sides), ..., the icosagon (20 sides), etc.

Regular polygons have interesting properties. For example, the centroid[2] of a regular polygon coincides with the center of the circle circumscribing the polygon. Moreover, the moment of inertia of the contour, a 2×2 symmetric and positive-definite matrix,[3] is proportional to the identity matrix. These properties make regular polygons attractive for the design of planar biaxial accelerometers, like the one depicted in Fig. 3.12. In the figure, the design of a biaxial accelerometer is shown, to be cut out of a silicon wafer, with a monolithic structure. This accelerometer is intended to measure the two components of the acceleration of the "proofmass"[4] that lie in its plane. The instrument consists of one *frame*, to be fixed to a *carrier*, whose acceleration is to be measured. A *proofmass* is suspended from the frame by means of articulated limbs. The joints of the limbs are produced by removing material around a point that is to function as the center of a *pin joint*. This type of *monolithic mechanism* is known in the art as a *compliant mechanism*.

2.4 Quadratic Curves: Conics

Quadratic curves, or conics, are the simplest of all 2D curves. Conics are used extensively in CAD work and geometric modeling.

In the most general sense, conics are curves formed by the intersection of a plane with a *right circular cone*.[5] The relative inclination of the plane with respect to the cone axis determines the conic produced: circle, ellipse, parabola, or hyperbola, as shown in Fig. 2.2.

Conics are commonly described in *implicit form* by the quadratic equation

$$Ax^2 + By^2 + 2Cxy + 2Dx + 2Ey + F = 0 \tag{2.15}$$

where (x, y) are the coordinates of an arbitrary point of the curve, and A, B, C, D, E, F are the coefficients characterizing the type of conic at hand.

In array form, and using homogeneous coordinates, this equation can be written as

$$\boldsymbol{p}^T \boldsymbol{R} \boldsymbol{p} = 0 \tag{2.16}$$

[2]See http://mathworld.wolfram.com/GeometricCentroid.html.

[3]Its eigenvalues (http://mathworld.wolfram.com/Eigenvalue.html) are positive.

[4]The proofmass is the triangular plate in the same figure.

[5]In a right circular cone, the intersections of all planes normal to the axis $\mathcal{A}$ are circles.

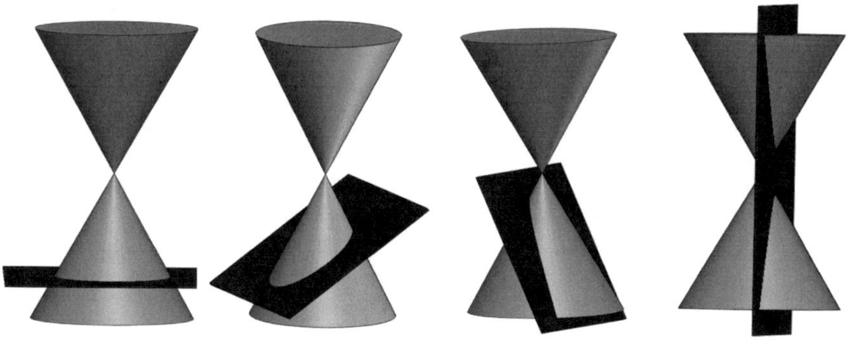

Fig. 2.2 Conics generated as the intersection of a right circular cone with a plane

where

$$\boldsymbol{p} = \begin{bmatrix} x \\ y \\ 1 \end{bmatrix} \quad \boldsymbol{R} = \begin{bmatrix} A & C & D \\ C & B & E \\ D & E & F \end{bmatrix} \tag{2.17}$$

Notice that matrix $\boldsymbol{R}$ is *symmetric*. In fact, it couldn't be otherwise because it is associated with a *quadratic form*, the one on the left-hand side of Eq. (2.15). As the reader can verify, if a skew-symmetric matrix $\boldsymbol{S}$ were added to $\boldsymbol{R}$ in the same equation, the contribution of $\boldsymbol{S}$ to the quadratic form would ineluctably vanish. Essentially, there are three different conic sections, the circle being a particular case of the ellipse. All three distinct sections are distinguished by the relationships among the six foregoing coefficients, as we shall describe below. In this description, the circle and the ellipse are discussed separately, while bearing in mind the foregoing statement.

2.4.1 The Circle

A circle is a *geometric primitive*, defined as the *locus* of the points equidistant from one point, the center of the circle. A circle is created when a plane passes either through a right circular cone or a cylinder, and is perpendicular to the axis of the cone (or cylinder, as the case may be), as shown in Fig. 2.2. Circles and their arcs are used extensively in engineering design, in particular, for the design of mechanical parts, be these made of metal, polymer, concrete, wood, or other construction materials.

Algebraic Representation of the Circle

Implicit Representation: The *implicit representation* of a circle with center at $C(x_1,\ y_1)$ and radius r is

$$(x - x_1)^2 + (y - y_1)^2 = r^2 \tag{2.18}$$

If the center is located at the origin (0, 0), the above equation simplifies to

$$x^2 + y^2 = r^2 \tag{2.19}$$

The *parametric representation* of a circle centered at the origin is

$$x = r\cos\theta, \quad y = r\sin\theta \tag{2.20}$$

where the *parameter* θ is the angle made by the position vector of coordinates (x, y) with the X-axis. If the center is not the origin, but a point of coordinates (x_1, y_1), then the parametric equation becomes

$$x = x_1 + r\cos\theta, \quad y = y_1 + r\sin\theta \tag{2.21}$$

Finally, we obtain the array representation of the circle defined in Eq. (2.21), by means of *homogeneous coordinates*, in the form

$$\boldsymbol{p}^T \boldsymbol{R} \boldsymbol{p} = \begin{bmatrix} x & y & 1 \end{bmatrix} \begin{bmatrix} 1 & 0 & D \\ 0 & 1 & E \\ D & E & F \end{bmatrix} \begin{bmatrix} x \\ y \\ 1 \end{bmatrix} = 0 \tag{2.22}$$

2.4.2 The Ellipse

Definition

An ellipse is a curve created when a plane intersects a right circular cone at an acute angle β, with the cone axis, greater than the acute angle α between the axis and the cone elements, as shown in Fig. 2.3.

An ellipse can be defined alternatively as the locus of all points in a plane whose distances from two fixed points F_1, F_2, the ellipse *foci*, in the plane, add to a constant,[6] as illustrated in Fig. 2.4.

An ellipse has two axes of symmetry that intersect at a point defined as the *center* of the ellipse, one being the *focal* axis, which contains the two foci; the other is normal to the focal axis and passes through the ellipse center C, as shown in Fig. 2.4. From the above definition, then

$$\overline{PF_1} + \overline{PF_2} = \text{const} \tag{2.23}$$

[6] On a practical note, an ellipse can be quickly constructed using a pencil whose tip is used to keep one string taut, whose two ends are, in turn, attached to the two foci of the desired ellipse.

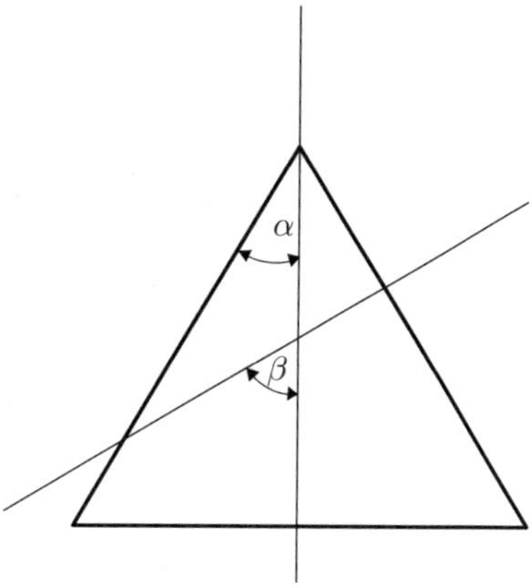

Fig. 2.3 The ellipse as a section of a right circular cone

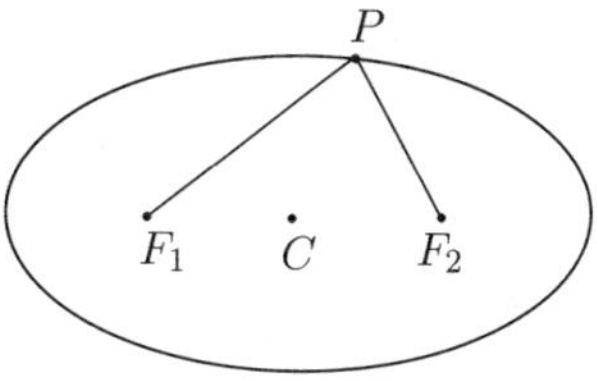

Fig. 2.4 The ellipse obtained by means of a taut string of length $\overline{PF_1} + \overline{PF_1}$, as per Eq. (2.23)

The **major axis** of an ellipse is the longest line segment included in the ellipse and passes through both foci. The **minor axis** is the perpendicular bisector of the major axis. The latter can also be defined as the line segment between two points of the ellipse that passes through the center and is of *minimum length*.

A circle viewed at an angle other than 90° (normal) appears as an ellipse due to perspective, as we can see in Fig. 2.5.

The ellipse possesses a reflective property: light or sound emanating from one focus is reflected to the other, a property useful in the design of some types of optical and auditory equipment. Whispering galleries, such as the Rotunda in the Capitol Building in Washington, D.C. and the Mormon Tabernacle in Salt Lake City, Utah, were designed using elliptical ceilings.

In a whispering gallery, low-intensity sound emanating from one focus is clearly audible at the other focus, but inaudible at any other points, hence the qualifier of *gallery*.

Examples of ellipses from the real world can be observed if

- A cylindrical glass partly filled with water is tilted; the free surface of the liquid will acquire an elliptical shape, as seen in Fig. 2.6.
- In the seventeenth century, Johannes Kepler discovered that each planet travels around the Sun in an elliptical orbit, with the Sun at one of its foci.
- We can also cite classical atomic theory: the electrons of an atom move in an approximately elliptical orbit, with the nucleus at one focus.

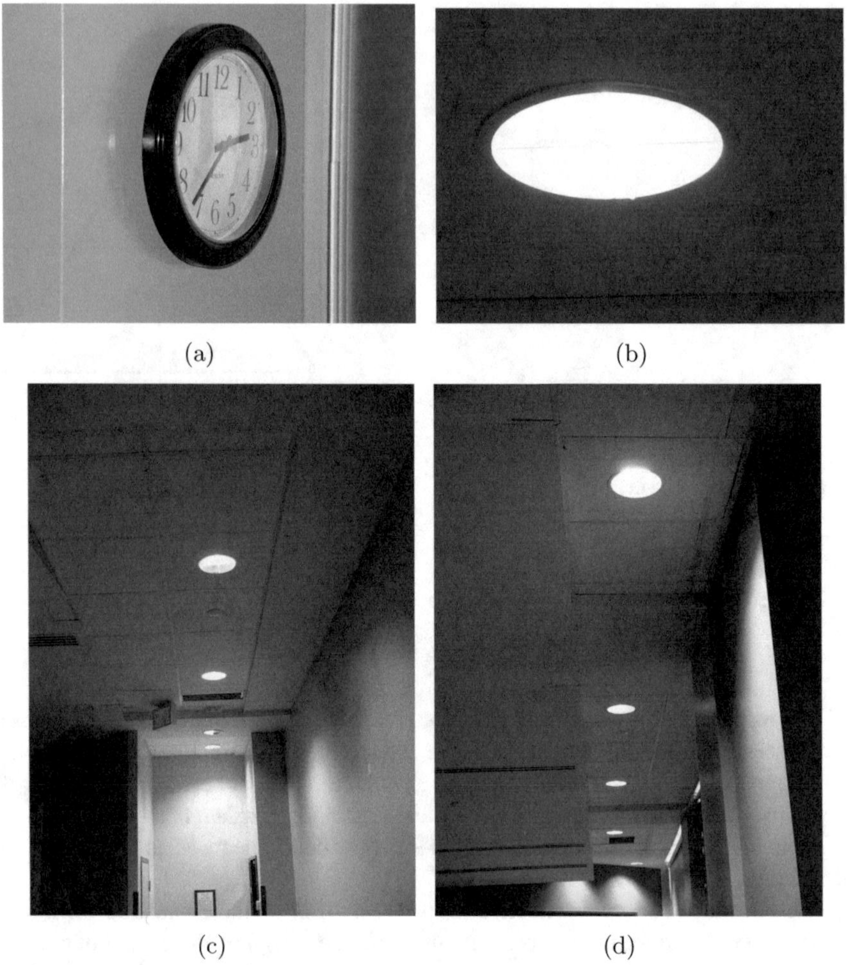

Fig. 2.5 Circles viewed as ellipses: **a** a Wall clock; **b** a close-up image of an aisle light; **c** and **d** aisle lights (Photos courtesy of V. Chopra, Ph.D.)

Algebraic Representation of the Ellipse

The *implicit representation* of an ellipse centered at the origin is given below in *canonical form*:

$$\frac{x^2}{a^2} + \frac{y^2}{b^2} = 1 \tag{2.24}$$

where the axes of this ellipse are assumed to coincide with the coordinate axes, as per Fig. 2.7. The constants a and b indicate the *semiaxis lengths*.

Note When the two axes are of the same length, the ellipse reduces to a circle.

Fig. 2.6 Tilting a glass of water (Photo courtesy of V. Chopra, Ph.D.)

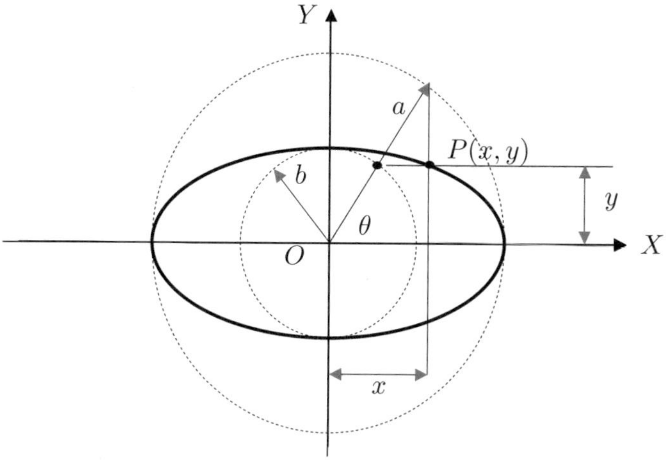

Fig. 2.7 The parametric representation of the ellipse

As illustrated in Fig. 2.7, the coordinates of a point of the ellipse are given by

$$x = a\cos\theta, \quad y = b\sin\theta \tag{2.25}$$

where θ is the parameter, while a and b are the semiaxis lengths, the axes of the ellipse coinciding with the coordinate axes, thereby obtaining the *parametric representation of the ellipse*.

Finally, we obtain the array form of the general ellipse equation in terms of the *homogeneous coordinates* of one of its points:

$$\boldsymbol{p}^T\boldsymbol{R}\boldsymbol{p} = \begin{bmatrix} x & y & 1 \end{bmatrix} \begin{bmatrix} A & C & D \\ C & B & E \\ D & E & F \end{bmatrix} \begin{bmatrix} x \\ y \\ 1 \end{bmatrix} = 0 \tag{2.26}$$

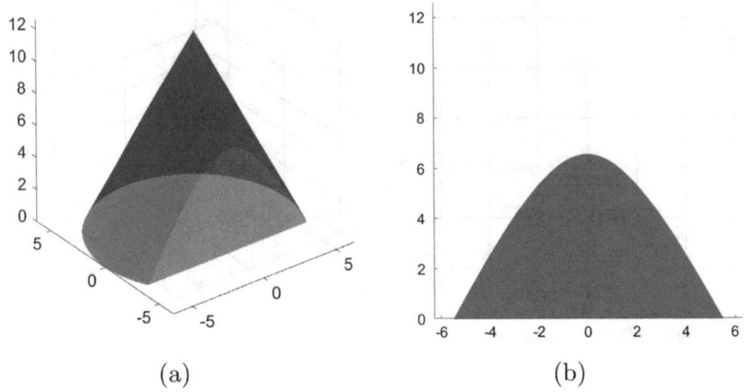

Fig. 2.8 A parabola determined by the intersection of a right circular cone with a plane parallel to an element of the cone: **a** general view; **b** normal view of the intersection curve

where A and B have the same sign.

2.4.3 *The Parabola*

Definition

A *parabola* is the curve created when a *right circular cone* is cut by a plane that is not normal to the cone axis *and* intersects the cone at only one nappe.[7] One instance is depicted in Fig. 2.8.

A parabola can be defined alternatively as the locus of the points in a plane that are equidistant from a given fixed point, called the *focus*, and a fixed line, called, in turn, the *directrix*, as depicted in Fig. 2.9.

Parabolas are quite useful in the design of engineering equipment due to a unique reflective property: Rays that originate at the focus of a parabola are reflected out of the parabola in a direction parallel to the axis. Conversely, rays entering the parabola parallel to the axis are reflected to the focus. Parabolas are thus used in the design of mirrors for telescopes, reflective mirrors for lights, cams for uniform acceleration, weightless flight trajectories, antennas for radar systems, arches for bridges, and field microphones commonly seen on the sidelines of football stadiums.

Parabolas can also be found in many other places:

[7] A right circular cone is defined as the surface generated by a line $\mathcal{L}$ that turns around a fixed line $\mathcal{A}$, the *cone axis*, while pivoting about a point P of $\mathcal{A}$ and making a constant angle with $\mathcal{A}$—See Fig. 2.12 and Sect. 3.3.2. A cone thus has two *nappes*.

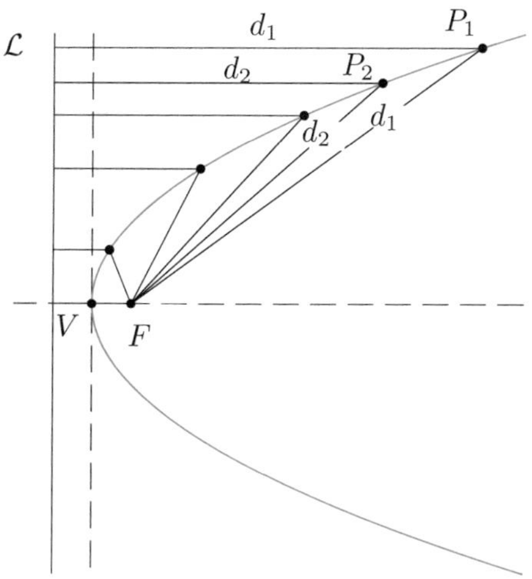

Fig. 2.9 The parabola defined as the locus of points equidistant from a focus F and a directrix $\mathcal{L}$

- One of nature's best known approximations to parabolas is the path taken by a body projected upward and obliquely to the pull of gravity, as in the trajectory of a golf ball.[8]
- A parabolic trajectory is also approximated by water emanating from a spout at a drinking fountain. Each molecule of water follows, approximately, a parabolic path, thus providing a picture of the curve, as shown in Fig. 2.10.
- In the design of communications equipment, antennas are often used to collect radio waves and light from a variety of *distant*[9] sources. The parabolic nature of the antenna allows it to collect and direct the signal at the focal point.
- A cable subjected to a uniformly distributed load along a horizontal direction adopts a parabolic form, an example being the *Golden Gate Bridge* in San Francisco, California, displayed in Fig. 2.11.

Algebraic Representation of the Parabola

Explicit Representation: The *canonical form* of a parabola is obtained when the curve passes through the origin and its focal axis coincides with the Y-axis, as given below in *explicit* form, with y as a function of x:

$$y = \frac{1}{4a}x^2 \tag{2.27}$$

where the Y-axis is, apparently, an axis of symmetry of the foregoing parabola, whereas a is a constant.

[8]This trajectory is not *exactly* a parabola because of the drag effect of the air. Golf balls on the moon would describe exact parabolas.

[9]The qualifier is intended to denote emitting sources located a distance from the focus much larger (several orders of magnitude) than the antenna focal distance, $\overline{VF}$ in Fig. 2.9.

Fig. 2.10 Trajectory of water ejected from a waterspout (Photo courtesy of V. Chopra, Ph.D.)

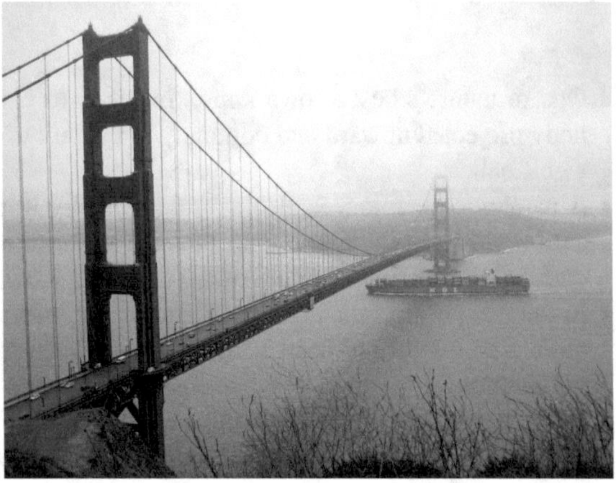

Fig. 2.11 The Golden Gate Bridge (Photo courtesy of V. Chopra, Ph.D.)

A more general representation of the parabola is given by

$$y = ax^2 + bx + c \tag{2.28}$$

where a, b, and c are constants.

Parametric Representation: The canonical form of Eq. (2.27) is transformed into parametric form, using the parameter t:

$$x = 2at, \quad y = at^2 \tag{2.29}$$

The center of a ball thrown with a horizontal velocity u_0 and a vertical velocity v_0 from a point of coordinates (0, h) describes a parabolic trajectory if air drag is negligible. The trajectory is given in parametric form by

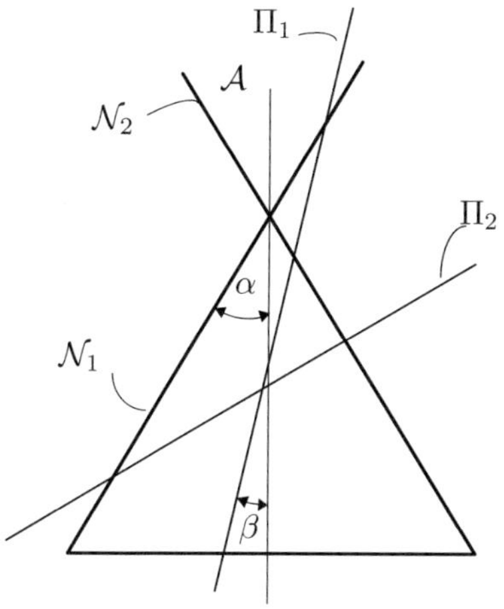

Fig. 2.12 The hyperbola generated by a plane cutting a right circular cone at an angle $\beta < \alpha$, α being the angle made by the elements with the cone axis

$$x = u_0 t, \quad y = h + v_0 t - \frac{1}{2} g t^2 \tag{2.30}$$

where g is the gravity acceleration, $9.8\,\mathrm{m/s}^2$, with one decimal and measured at the sea level. Since the gravity constant g is smaller on the Moon than on the Earth and the third term in the right-hand side of the second equation in Eq. (2.30) is negative definite, golf balls on the Moon would go farther when hit by the same player and with the same force. The trajectory would thus still be a parabola, but "more open" that its counterpart on the Earth.

Finally, the array form of the parabola is given by

$$\boldsymbol{p}^T \boldsymbol{R} \boldsymbol{p} = \begin{bmatrix} x & y & 1 \end{bmatrix} \begin{bmatrix} A & C & D \\ C & B & E \\ D & E & F \end{bmatrix} \begin{bmatrix} x \\ y \\ 1 \end{bmatrix} = 0 \tag{2.31}$$

where $AB - C^2 = 0$.

2.4.4 The Hyperbola

Any cone, in particular a *right circular cone*, has two nappes, as shown in Fig. 2.2. The hyperbola is the curve obtained when such a cone is cut by a plane Π_1 at its two nappes, $\mathcal{N}_1$ and $\mathcal{N}_2$, as shown in Fig. 2.12. Notice that plane Π_2, in the same figure, cuts only $\mathcal{N}_1$.

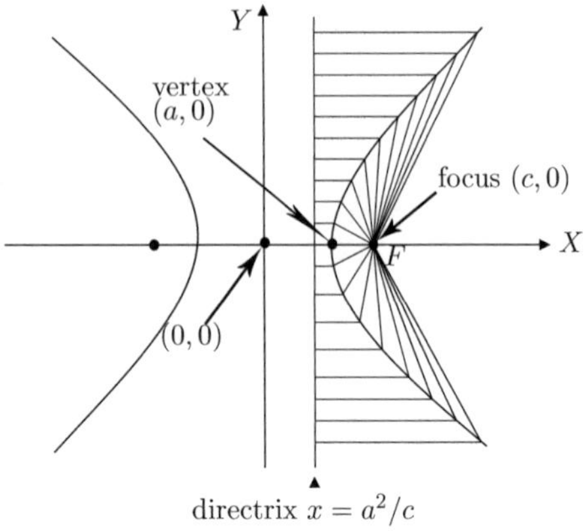

Fig. 2.13 The hyperbola as the locus of points whose distance from a focus F is proportional to the horizontal distance from a line, the directrix

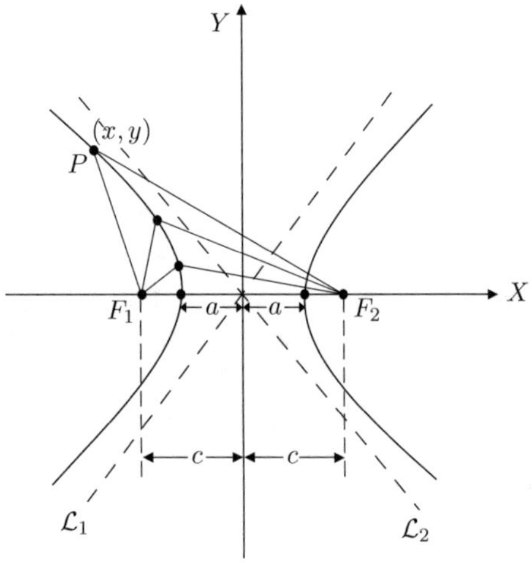

Fig. 2.14 The hyperbola as the locus of points P obeying the property $\overline{PF_2} - \overline{PF_1} =$ constant

A hyperbola can be defined alternatively as the locus of all points in a plane whose distance from a focus F is proportional to the horizontal distance from a vertical line, the directrix, as depicted in Fig. 2.13.

One more alternative definition of the hyperbola follows: *the locus of all points in a plane whose distance from two fixed points, called the foci (lying in the plane as well), have a constant difference*, as illustrated in Fig. 2.14.

An interesting *reflective property* of the hyperbola is illustrated in Fig. 2.15: lines F_1P and F_2P make equal angles θ with the tangent to the curve at P.

Hyperbolas are also found in many places:

- When alpha particles are shot toward the nucleus of an atom, they are repulsed away from the nucleus along hyperbolic paths.
- In astronomy, a comet follows a hyperbolic path, with the sun as one of its foci.

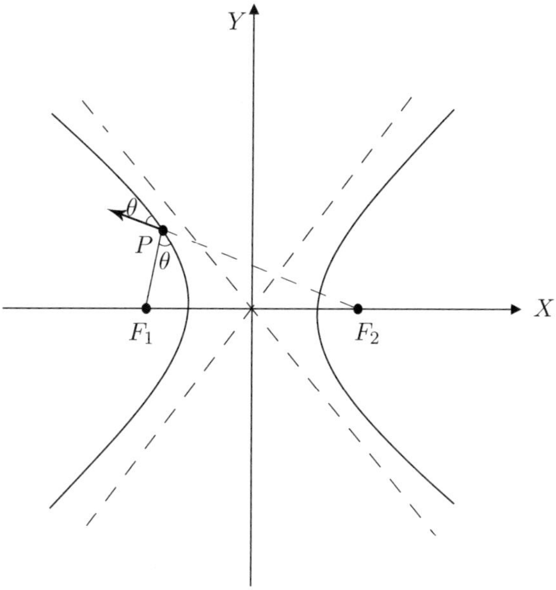

Fig. 2.15 A reflective property of the hyperbola: lines $\overline{F_1P}$ and $\overline{F_2P}$ make equal angles with the tangent at P

Fig. 2.16 The two branches of the hyperbola determined by the corresponding rims of a cylindrical shade, as projected on a wall (photo courtesy of Ms. A.-M. Esnos)

- The light rays stemming from a light bulb inside a lampshade define, with the rim of the shade, a cone of light. Assuming that the rim is a circle lying in a vertical plane, the axis of the cone is horizontal. Its intersection with a vertical wall traces a hyperbola, as shown in Fig. 2.16.

Algebraic Representation of the Hyperbola

Implicit Representation: The canonical representation of the hyperbola is given by

$$\frac{x^2}{a^2} - \frac{y^2}{b^2} = 1, \quad b^2 = c^2 - a^2 \tag{2.32}$$

its parameters a and c being illustrated in Fig. 2.14, in which the hyperbola axes coincide with the coordinate axes. The definitions of major and minor axes for the hyperbola follow those given for the ellipse. The X-axis intersects the curve at two points, $(0, -a)$ and $(0, a)$, but the Y-axis does not intersect the curve at all.

The parametric representation of the hyperbola with focal axis coincident with the X-axis bears a striking resemblance to that of the ellipse:

$$x = a \sinh\theta, \quad y = b\cosh\theta \tag{2.33}$$

where $\cosh\theta$ and $\sinh\theta$ are the *hyperbolic functions*, *hyperbolic cosine* and *hyperbolic sine*, of the real variable θ, defined as

$$\cosh x = \frac{\exp(x) + \exp(-x)}{2}, \quad \sinh x = \frac{\exp(x) - \exp(-x)}{2} \tag{2.34}$$

Therefore, the only difference with the counterpart equation of the ellipse lies in that the latter includes the *trigonometric* functions $\cos\theta$ and $\sin\theta$.

Furthermore, notice that the canonical form of the equation of the hyperbola can be cast in the form

$$(ay + bx)(ay - bx) = a^2b^2 \tag{2.35}$$

which is obviously violated by all points lying on the lines

$$\mathcal{L}_1\colon\ ay + bx = 0, \quad \mathcal{L}_2\colon\ ay - bx = 0 \tag{2.36}$$

Lines $\mathcal{L}_1$ and $\mathcal{L}_2$, illustrated in Fig. 2.14, are called the *asymptotes*—Greek: *a*, negation; *symptotos*, falling together—of the hyperbola.

Finally, we obtain the array form of the generalized hyperbola:

$$\boldsymbol{p}^T\boldsymbol{R}\boldsymbol{p} = \begin{bmatrix} x & y & 1 \end{bmatrix} \begin{bmatrix} A & C & D \\ C & B & E \\ D & E & F \end{bmatrix} \begin{bmatrix} x \\ y \\ 1 \end{bmatrix} = 0 \tag{2.37}$$

where A and B bear opposite signs.

Summary

In general, all types of conics have many engineering and architectural applications; they can be utilized in combination with one another to create intricate machines and architectural structures.

Using the generalized implicit form of the general equation for the conics, upon expansion of the intermediate side of Eq. (2.37), the generic equation is obtained as follows:

$$Ax^2 + By^2 + 2Cxy + 2Dx + 2Ey + F = 0 \tag{2.38}$$

We can now identify the type of conic simply by observing the sign of the *discriminant* of the above polynomial, namely,

$$\Delta = AB - C^2 \tag{2.39}$$

which is nothing but the *determinant* of the top-left 2×2 block of the 3×3 matrix $\boldsymbol{R}$ in Eq. (2.17).

Thus, we can identify the type of conic:

If $\Delta > 0$, the **general equation represents an ellipse.**
If $\Delta = 0$, **the general equation represents a parabola.**
If $\Delta < 0$, **the general equation represents a hyperbola.**

Notice that a circle is identified by the condition $\Delta > 0$ *and* the relation $A = B$.

2.5 Higher Order Algebraic Curves

Let us consider a curve $\mathcal{C}$ described by an implicit equation $f(x, y) = 0$, where function $f(x, y)$ involves products $x^i y^j$, for i and j any natural numbers. Such a curve, described by a *bivariate polynomial* in x and y, is termed an *algebraic curve*. The order, or degree, $d_{\mathcal{C}}$ of an algebraic curve $\mathcal{C}$ is defined as

$$d_{\mathcal{C}} \equiv \max_{i,j}\{i + j\} \tag{2.40}$$

The conics are, thus, second-order curves. Any algebraic curve with $d_{\mathcal{C}} > 2$ will be termed in this book a *higher order curve*. Now we have the result below.

Fact *An algebraic curve of degree d intersects a line at d points at most.*

Proof The proof is straightforward. Consider the line $\mathcal{L}$ given by

$$\mathcal{L}: \quad Ax + By + C = 0, \quad AB \neq 0 \tag{2.41}$$

Under the assumption that $B \neq 0$—if B turns out to vanish, then we can solve for x as $-C/A$, a constant value, and proceed in a slightly different, although similar way. We can thus solve for y in terms of x, namely,

$$y = -\frac{A}{B}x - \frac{C}{B}$$

When the above expression is substituted into $f(x, y) = 0$, a *monovariate polynomial* equation $P(x) = 0$, of degree d, is obtained. Now, the equation thus resulting has exactly d roots, whether real or complex, with complex roots occurring in conjugate pairs. Each such real root thus defines one intersection of $\mathcal{L}$ with $\mathcal{C}$, thereby proving that the line intersects the curve at d points at most.

A special class of algebraic curves that finds ample applications in design are *Lamé curves*, thus named after the French mathematician Gabriel Lamé (1795–1870), who first proposed them. These are m-order curves, endowed with interesting properties, that take the simple forms

$$f(x, y) = x^m + y^m - 1 = 0 \tag{2.42}$$

These curves are plotted in Fig. 2.17 for $m = 2, \ldots, 7$. Properties of these curves are given below:

- Lamé curves of even degree are closed and symmetric with respect to the x- and y-axes;
- Lamé curves of odd degree (i) are open and symmetric with respect to a line passing through the origin and making an angle of 45° with the X-axis, and (ii) have an asymptote, the line passing through the origin and making an angle of 135° with the same axis.
- Moreover, the curvature of the Lamé curves of both even and odd degrees vanishes at the intersections with the coordinate axes, except for $m = 2$, in which case the curvature is constantly equal to unity.

While the Lamé curves defined above are *normalized*, in that the coefficients of x and y are unity, scalings in the directions of the coordinate axes are possible by means of an *affine transformation*, as outlined in Sect. 4.1.1.

It is noteworthy that the Lamé curves for $m = 3$, 5, and 7 in Fig. 2.17 exhibit an open-ended shape, i.e., these curves extend infinitely toward the second and the fourth quadrants, its points at distances $d \gg 1$ from the origin approaching its asymptote. Similar to the case of the hyperbola, we can find the asymptote of the cubic Lamé curve upon first writing Eq. (2.42), for $m = 3$, in the form

$$x^3 + y^3 = 1$$

Next, we factor the left-hand side of the above equation into a linear and a quadratic factor, namely,

$$(x + y)(x^2 - xy + y^2) = 1 \tag{2.43}$$

Apparently, the set of points on the line $x + y = 0$ never touches the curve, and hence this line is the asymptote of the curve. As a matter of fact, all odd-degree Lamé curves obeying an equation of the form of Eq. (2.42) have the same asymptote.

The parametric representation of Lamé curves is now introduced: Let P be an arbitrary point on a Lamé curve of degree m and θ the angle between line OP and the X-axis, measured positive ccw. The Cartesian coordinates of P are now given by

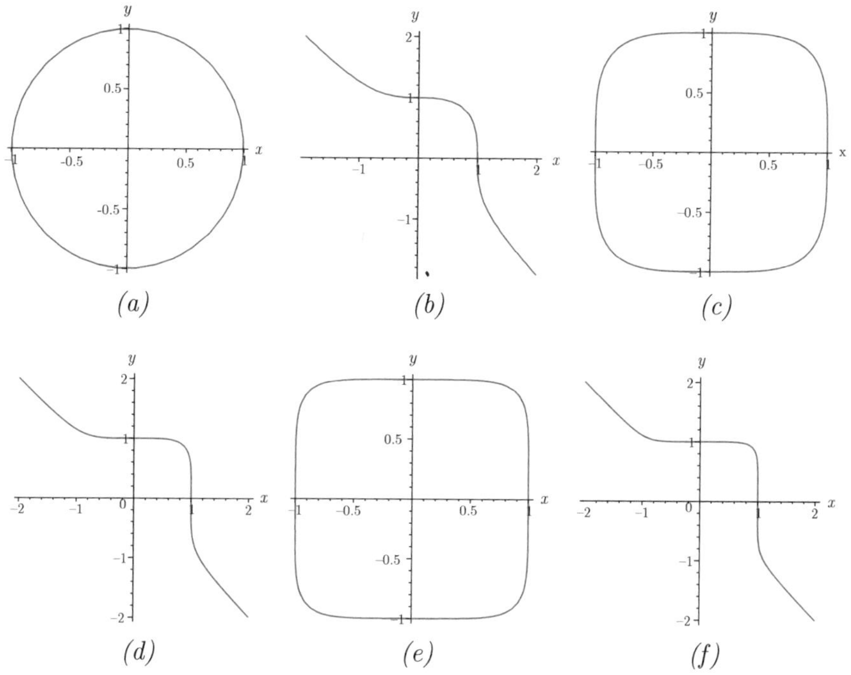

Fig. 2.17 Plots of the Lamé curves for $m = 2, \ldots, 7$

$$x(\theta) = \frac{1}{[1 + \tan^m(\theta)]^{1/m}}, \quad y(\theta) = \frac{\tan\theta}{[1 + \tan^m(\theta)]^{1/m}}, \quad 0 \le \theta < \frac{\pi}{2} \tag{2.44}$$

where the strict inequality is intended to avoid the *singularity*[10] of the $\tan(\cdot)$ function.

Lamé curves are ideal to provide blending of flat features of mechanical or architectural objects under design. Indeed, the standard blending surface is produced out of an arc of a circle, but this practice brings about problems of stress concentration in structural elements. Indeed, according to structural mechanics, failure in convex regions of a structure occurs at points of either high curvature values or curvature discontinuities.[11] Obviously, blending a line with a circle at a point of tangency, the finite curvature of the circle—the reciprocal of its radius—undergoes a sudden change to 0, the reciprocal of the "radius of curvature" of the line. The latter can be regarded as a circle of infinite radius. Here is where Lamé curves come to the rescue, as illustrated below with an example of mechanical design. An application of Lamé curves to mechanical design along these lines is the mechanism housing included in Figs. 1.2 and 1.4.

[10]Points at which the value of a function becomes unbounded are termed *singular*.

[11]See: Teng et al. (2008) and the references therein.

Fig. 2.18 A close-up of the mechanism cover of Fig. 1.4

A close-up of a critical region of the mechanism housing of Figs. 1.2 and 1.4 is shown in Fig. 2.18. This region blends the bottom and the top parts of the mechanism housing with the blue, in Fig. 1.4 blocks that receive the roller bearings.[12] The critical region in question was designed with a *smooth blending*, to avoid both high curvature values and curvature discontinuities. The inner surface of the region in question was generated by means of a sixth-order Lamé curve, the outer surface by means of an eighth-order Lamé curve. The reason for the different orders lies in that, as made apparent by the plots of the Lamé curves of Fig. 2.17, the maximum value of the curvature occurs when $x = y$. Moreover, the higher the value of m, the "sharper" the curve becomes at the value of maximum curvature. The reader can verify this statement by evaluating the curvature formula of Eq. (2.42) at $x = y$—or at $x = -y$ as well—for even values of m in the same formula. According to the *theory of failure*, this occurs at concave regions, and hence, if the cover fails, it will do so at interior points. Hence, the rationale to adopt a lower order curve for the inner surface.

The general (implicit) equation representing a cubic curve takes the form

$$f(x, y) \equiv A_{30}x^3 + A_{21}x^2y + A_{12}xy^2 + A_{03}y^3 + A_{20}x^2 + A_{11}xy + A_{02}y^2 + A_{10}x + A_{01}y + A_{00} = 0 \tag{2.45}$$

Finding the asymptote of a general cubic may be more challenging, for this requires finding two factors, one linear and one quadratic, of $f(x, y)$, which is not a simple task. More general procedures are available for computing asymptotes of curves, but these fall beyond the scope of this book, and will hence not be pursued.

2.6 Non-algebraic Curves

Curves described by the implicit function $f(x, y) = 0$, in which $f(x, y)$ does not become a polynomial in x upon fixing y to a constant value, or the other way around, upon fixing x to a constant value, are called *non-algebraic*. In this case, the number of intersections of the curve with a line may be infinite. Examples of non-algebraic

[12]Only the top part is shown in Fig. 2.18.

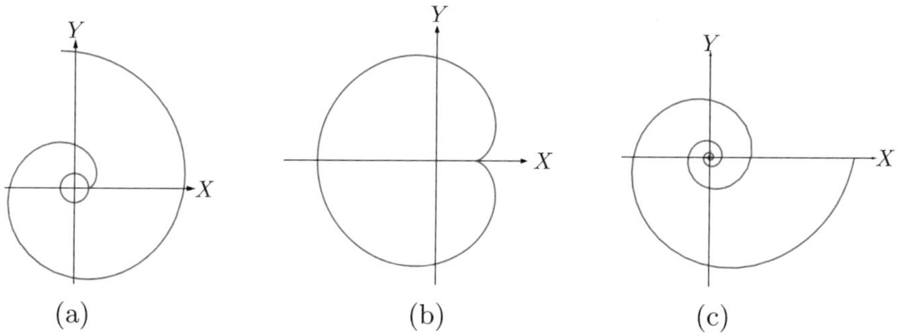

Fig. 2.19 Examples of non-algebraic curves: **a** the circle-involute; **b** the epicycloid; and **c** the logarithmic spiral

curves abound. Of special interest in design are the *circle-involute*, the *epicycloid* and the *logarithmic spiral*. Instances of these curves are included in Fig. 2.19.

The *circle-involute* is obtained upon wrapping a string to a drum, and then unwrapping the string while keeping one of its ends fixed to the drum and maintaining the string taut as it unwraps. Its parametric equations are

$$x = a(\cos t + t \sin t), \quad y = a(\sin t - t \cos t) \tag{2.46}$$

The *epicycloid* is obtained as a circle of radius b rolls without slipping around a larger circle, of radius a. An arbitrary point on the circumference of radius b then traces the epicycloid. The parametric equations of the epicycloid are

$$x(t) = (a + b)\cos\phi - b\cos\left(\frac{a+b}{b}\phi\right), \quad y(t) = (a + b)\sin\phi - b\sin\left(\frac{a+b}{b}\phi\right) \tag{2.47}$$

The foregoing curves find applications in the design of *spur gears*, i.e., gears capable of transmitting motion between two shafts of parallel axes. The form of these gears is fully described by planar contours. The most frequent are gears with involute teeth, those with epicycloid teeth being less common. Furthermore, the logarithmic spiral can be used as an *auxiliary curve* to design gear-tooth profiles with the shape of a circle-involute.[13]

2.7 Free-Form Curves

Sometimes, design applications call for curves that cannot be represented, either in full or piecewise, by simple implicit functions of the form $f(x, y) = 0$. These are

---

[13]Figliolini et al. (2019).

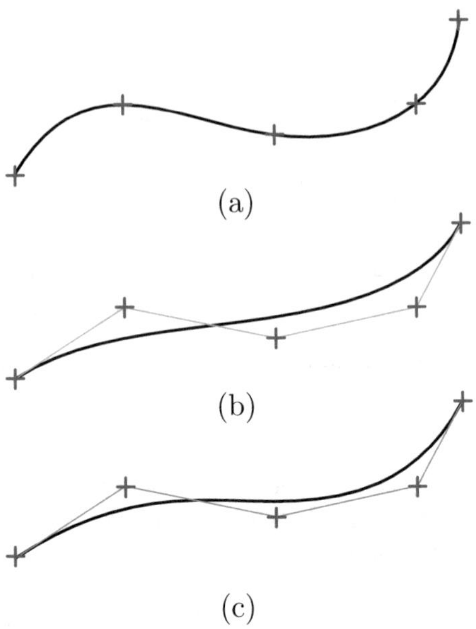

Fig. 2.20 Free-form curves: **a** spline; **b** Bézier; **c** B-spline

called *free-form curves*. The automobile industry uses many free-form curves in the design of the vehicle body. Other applications abound, mainly in connection with free-form surfaces, as outlined in Chap. 3.

Splines are among the most frequently used curves in the auto, aircraft, and ship-building industries. The cross section of an airplane wing or a ship hull is a spline curve. In addition, spline curves are commonly used to define the path of motion in computer animation. For CAD systems, three types of free-form curves have been developed: splines, Bézier curves, and B-spline curves, as depicted in Fig. 2.20.

Bézier curves were invented simultaneously by Paul de Casteljau at Citroën and Pierre E. Bézier at Renault around the late 50s and early 60s. However, Bézier was able to publish his work in several journals, thus bestowing his name on the newly created family of curves.

These curves can be described by sets of parametric equations, in which the x- and y-coordinates of the *control points* are computed as functions of a third variable, called a *parameter*, as in Eqs. (2.46) and (2.47).

The topic of free-form curves is rather advanced,[14] for which reason it is not pursued in this book. In the balance of the section, we expand on the concepts of curve continuity, as needed in engineering design.

In the spline displayed in Fig. 2.20a three intermediate points are shown, where four algebraic curves meet pairwise. At each of these points, the two curves are *forced* to share the point in question, which is termed G^0-continuity, with G standing for *geometric*, as opposed to *analytic* continuity, proper of curves representing functions,

[14]Mortenson (1985).

which is represented with C. With reference to the same figure, the two *blended curves* share not only one common point, but also one common tangent, which is termed G^1-continuity. If, furthermore, the two blended curves are forced to share the same *center of curvature*, and hence the same curvature, then, we speak of G^2-continuity. Higher order continuity is needed in some applications. In this book, however, we will not consider such special applications.

G^N-continuity, for $N = 0, 1, 2$, calls for some additional concepts, as outlined below. The grasping of these concepts is eased if we assume a *parametric representation* of the curve $\mathcal{C}$ of interest:

$$\mathcal{C}: \quad x = x(p), \quad y = y(p) \tag{2.48}$$

with p being the parameter, a length or an angle, that describes the curve.

The *tangent* to $\mathcal{C}$ at point $P(x, y)$ is described by the unit vector $\boldsymbol{e}_t$, computed as

$$\boldsymbol{e}_t = \frac{1}{\sqrt{x'(p)^2 + y'(p)^2}} \begin{bmatrix} x'(p) \\ y'(p) \end{bmatrix} \tag{2.49a}$$

or, if the position vector of P is $\boldsymbol{r} = [x(p), y(p)]^T$, then

$$\boldsymbol{e}_t = \frac{\boldsymbol{r}'(p)}{\|\boldsymbol{r}'(p)\|} \tag{2.49b}$$

The *curvature* κ of $\mathcal{C}$ at P is given, in turn, by

$$\kappa(p) = \frac{x'(p)y''(p) - y'(p)x''(p)}{[x'(p)^2 + y'(p)^2]^{3/2}} \tag{2.50a}$$

or, in terms of the position vector $\boldsymbol{r}$ and its derivatives w.r.t. p,

$$\kappa(p) = -\frac{\boldsymbol{r}'(p)^T \boldsymbol{E}\boldsymbol{r}''(p)}{\|\boldsymbol{r}'(p)\|^{3/2}} \tag{2.50b}$$

The curvature of the m-degree Lamé curve, in particular, takes its simplest form in the implicit representation of Eq. (2.42):

$$\kappa(x, y) = \frac{(m-1)(xy)^{m-2}(x^m + y^m)}{(x^{2m-2} + y^{2m-2})^{3/2}}. \tag{2.51}$$

Apparently, the curvature of the Lamé curves vanishes whenever x or y does, i.e., at the intersection of the curve with the coordinate axes. However, this statement is true as long as $m \neq 2$. When $m = 2$, the curve becomes a circle of unit radius, whose curvature is constant and equal to unity.

In order to illustrate the smoothness of the curvature distribution of Lamé curves, the curvature of the fourth-degree curve is plotted in Fig. 2.21.

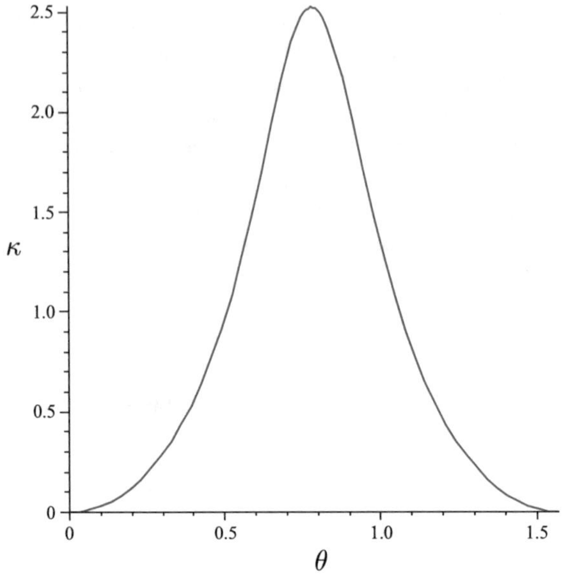

Fig. 2.21 Curvature distribution of the fourth-degree Lamé curve

In Fig. 2.22a, two line segments at right angles are blended with a circular arc of radius r. At the blending points, we have G^0- and G^1-continuities, but G^2 discontinuity, as the curvature changes suddenly from 0 on the line segment to $1/r$ on the arc, and vice versa. In Fig. 2.22b, in turn, we have the blending of two line segments with a fourth-degree Lamé curve. By virtue of the smooth curvature distribution of Lamé curves in general, a paradigm of which is shown in Fig. 2.21, the blending is G^0-, G^1-, and G^2-continuous. In Fig. 2.22c, we have one additional example of a smooth, non-algebraic curve that finds extensive applications in design: the *cycloid*. At the points lying on the X-axis, separated by a *period* a, the curve is apparently G^0- and G^1-continuous, but its curvature grows unbounded at these points, as the reader can verify upon application of the formula (2.50a) or (2.50b).

The parametric equations of the cycloid are

$$x = a(t - \sin t), \quad y = a(1 - \cos t) \tag{2.52}$$

Furthermore, the spherical counterparts of the circle-involute, the epicycloid, and the logarithmic spiral are well defined, although beyond the scope of the book. These find applications in the design of bevel gears, i.e., gears intended for the transmission of torque and motion between shafts with intersecting axes, normally at right angles.

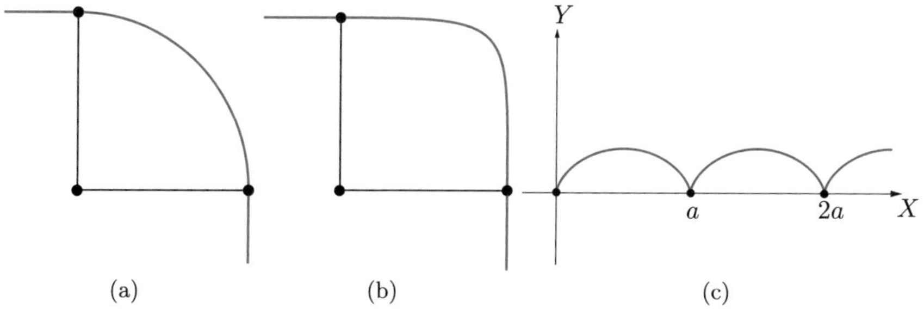

Fig. 2.22 Blending points with: **a** G^0- and G^1-continuity, but G^2-discontinuity; **b** G^0-, G^1-, and G^2-continuity; and **c** G^0- and G^1-continuity, but G^2-discontinuity

2.8 Properties of Planar, Bounded Figures

Any planar figure in the plane can be translated and rotated inside the plane, without changing its shape and dimensions—notice that shape can be preserved under a *uniform* scaling. In fact, any set of material points with this property is known in mechanics as a *rigid body*. In two dimensions, of course, we are interested in planar rigid bodies. The reader can visualize a planar rigid body as a slab of undeformable material with a uniform thickness that can move freely on a planar, horizontal surface, while all its points remain in contact with the surface.

Planar figures describe completely pieces frequently used in engineering and architecture, which exhibit a flat shape. The manufacturing of such pieces, produced of metals, can be realized by *milling* or *water-jet cutting*. An instance of such pieces, manufactured with water-jet cutting, is shown in Fig. 2.23. In the figure, we have, in fact, two identical pieces fabricated with this technology, fastened into one single piece, the latter being driven via a cable–pulley mechanism by two motors with parallel axes. The two identical pieces are parts of a robot intended for pick-and-place operations.[15]

A planar rigid body, or any rigid body for that matter, is defined as a *continuum set of points with one property: under whatever displacement the body undergoes, all its points remain equidistant*. One more important property of the rigid body is its *pose*, given by (i) the position of any of its points and (ii) the orientation of any of its lines with respect to a line fixed on its plane of motion. The orientation of the body is also known as its *attitude*.

[15]Belzile et al. (2020).

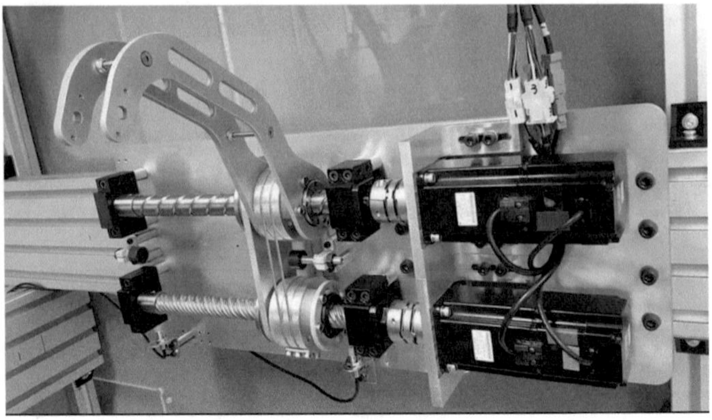

Fig. 2.23 Robotic arm manufactured with water-jet cutting (Photo courtesy of B. Belzile, Ph.D.)

2.9 Summary

The production of planar objects, i.e., objects defined by planar curves, is the aim of this chapter. The chapter starts with the basic concepts of planar geometry: points and lines, and then quickly goes into planar curves. The simplest of these are the conics, defined by second-degree *bivariate polynomials*. These curves are far from being found only in textbooks. The reader is alerted to the existence of such curves in the most common places, such as the house, the office, the school, or even bridges. The application of these curves to the production of engineered objects, such as parabolic antennae, solar mirrors, or whispering galleries, is brought to the limelight.

Conic sections, ubiquitous as they are, however, are not sufficient to produce designed objects to meet certain specifications. For example, a frequent task in civil and mechanical engineering design work requires to blend two coplanar line segments with a smooth curve. While the usual practice is to use an arc of a circle, structural engineers have found that this solution suffers from limitations. Indeed, mechanical failure occurs at points of a profile that exhibit "large" curvature values or, even worse, curvature discontinuities. Circles offer continuity of tangency, but they cannot offer continuity of curvature. Lamé curves come to the rescue, but these curves are rarely mentioned in CAD or elementary geometry textbooks. These curves are given due attention in this chapter. Lamé curves are simple *algebraic curves*, described by polynomials of degree three or higher. Other curves, termed *non-algebraic* because they are described by *transcendental functions*, such as the involute, the cardioid, the epicycloid, and the logarithmic spiral, were also introduced. These curves find applications in the design of gear profiles, among others. An introduction to Lamé curves and non-algebraic curves is provided. The chapter ends with an introduction to free-form curves, which find applications in the design of the aerodynamic contours of terrestrial vehicles and aircraft. A crash introduction into *splines* is provided at the end of the chapter, as such curves (i) find extensive applications in engineering design in industrial environments and (ii) are available in every modern CAD system.

2.10 Exercises

2.1 (a) For the conic below:

$$f = 3x^2 - 2xy - 14x + 3y^2 + 10y + 18 = 0$$

find its array representation of the form of Eq. (2.16), while providing numerical values for the entries $A, B, \ldots, F$, and sketch the conic in question. What is the type of this conic?

(b) Repeat the above exercise for the curve given by

$$g = 2x^2 - 6xy + 2x - 3y^2 + 2y + 1 = 0$$

(c) Repeat the same exercise for the curve given by

$$h = 2x^2 + 2\sqrt{6}xy + 2x + 3y^2 + 2y + 1 = 0$$

2.2 Using scientific software (Maple, Mathematica, or Matlab), plot the curves below:

(a) An ellipse with foci on the X-axis, the foci being symmetrically located with respect to the origin, a distance $d = 10$ units apart; moreover, the sum of the distances from any point P on the ellipse to the two foci is 20 units.
(b) A parabola with focus at the origin and directrix parallel to the X-axis, 5 units below this axis; and
(c) A hyperbola with foci located as described in item (a) above, the difference between the distances from any point on the hyperbola to the two foci being 8.

2.3 Three conics are given below:

(a) $x^2 - 2xy - 4 = 0$
(b) $xy - 1 = 0$
(c) $3x^2 + 2xy + 3y^2 + 10x - 2y + 10 = 0$

Identify each conic as circle, ellipse, hyperbola or parabola.

2.4 Three more conics are given below:

(a) $(x - 2)(y - 3) - 1 = 0$
(b) $(x + y + 1)(x - y - 1) + 1 = 0$
(c) $(x + 2)(y + 3) + 1 = 0$

Identify each conic as circle, ellipse, hyperbola, or parabola.

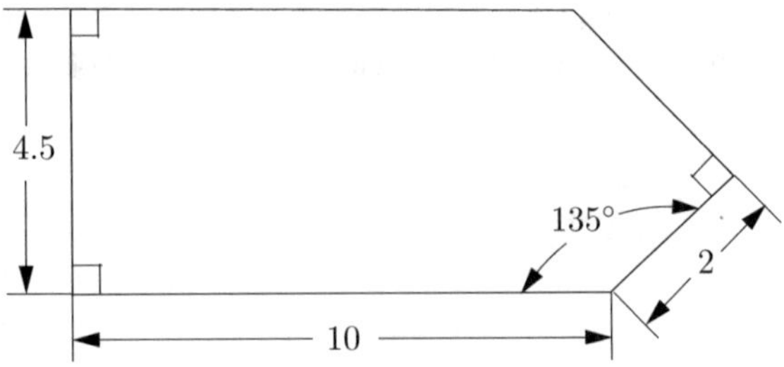

Fig. 2.24 An irregular polygon

2.5 Using scientific software, construct the irregular polygon shown in Fig. 2.24, using the given dimensions, in mm.

2.6 With a pencil, free-hand sketch the bracket shown in Fig. 2.25 on a letter-size sheet. Lengths are in mm.

2.7 Using scientific software, plot an ellipse of center located at point $C(3, 4)$, with the focal axis parallel to the X-axis and the foci a distance of 5 units from C, if we know that the distances of a point P of the ellipse to each of the foci are 10 and 5 units. Then, plot a hyperbola with the same foci and a point Q of the hyperbola lying a distance of 10 and 5 units from each focus. The use of the procedure to plot *implicit* curves is recommended.

2.8 (i) Using scientific software, trace the ellipse defined by

$$\frac{x^2}{4} + \frac{y^2}{2} = 1,$$

and find the positions of its foci.
(ii) Plot one of the parabolas tangent to the previous ellipse, with axis at 45° with the X-axis, and intersecting the Y-axis a distance c from the origin, with c defined as the focal distance of the ellipse.

2.9 Let a line $\mathcal{L}$ go through the point Q of position vector $\boldsymbol{q} = [\,2\ 3\,]^T$ and have direction $\boldsymbol{v} = [\,3\ -4\,]^T$. Further, let $\mathcal{H}$ be a hyperbola with foci on the X-axis, both located a distance $c = 5$ from the origin, and with the difference between the distances from a point of $\mathcal{H}$ to the foci equal to 6. Trace the two previous objects with scientific software and find any intersection(s) between them.

2.10 A structural plate in a heavy-duty machine requires a rectangular opening of 500 mm × 300 mm, in order to provide access for servicing, maintenance, and repair. Provide a drawing, using scientific software, of the opening, with a fillet that blends the two edges of each corner with a fourth-order Lamé curve. The tangency points of the curve and the edges should lie a distance of 10 mm from the corners. Include a zoom-in of a typical corner.

2.11 Produce a table of numerical values of the coordinates of a set of points on a cubic Lamé curve of the form

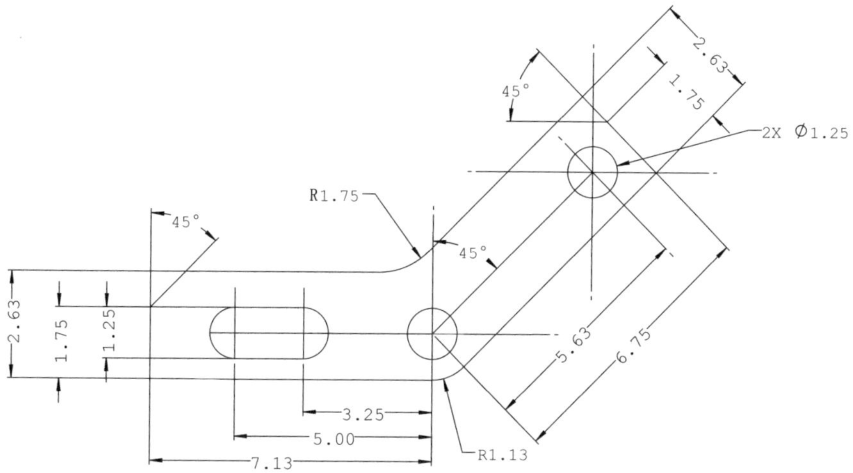

Fig. 2.25 An angle bracket

$$x^3 + y^3 = 1$$

in the first quadrant. Moreover, the position vectors of every pair of neighboring points should make an angle $\Delta\theta = 3°$. *Hint: Trace lines from the origin and intersecting the curve at a set of points* $\{P_i\}_1^m$ *at corresponding distances* $\{d_i\}_1^m$ *from the origin. Each distance* d_i *can be obtained as the real root of a cubic equation.*

2.12 Find the numerical value of the peak curvature of the plot of Fig. 2.21. *Hint: the maximum curvature of even-degree Lamé curves apparently occurs at* $x = \pm y$.

2.13 Show that the radius of curvature—the reciprocal of the curvature—of the cycloid, with parametric equations given by Eq. (2.52), at values $t = 2\pi k$, for natural values of k, vanishes.

References

Belzile B, Eskandary PK, Angeles J (2020) Workspace determination and feedback control of a pick-and-place parallel robot: analysis and experiments. IEEE Robot Automat Lett 5(1):40–47

Figliolini G, Stachel H, Angeles J (2019) Kinematic properties of planar and spherical logarithmic spirals: applications to the synthesis of involute tooth profiles. Mech Mach Theory 136:14–26

Mortenson ME (1985) Geometric Modeling. Wiley, New York

Teng CP, Bai S, Angeles J (2008) Shape synthesis in mechanical design. Acta Polytech 47(6):56–62

Chapter 3
3D Objects

This chapter is devoted to objects defined by sets of points lying in 3D space, i.e., located in a reference frame $X,\ Y,\ Z$. The counterpart of points and lines are similar to those studied in Chap. 2. Lines are defined as in Chap. 2, and given by one point and one unit vector, which now has three components. A major difference between the current chapter and Chap. 2 lies in that a rigid body is now a solid object, delimited by a boundary that is a surface. Moreover, one novel concept is the *distance between two skew lines*, a concept that did not arise in Chap. 2. Furthermore, because of their extensive applications in design, *quadrics*, i.e., three-dimensional surfaces defined by a *smooth* quadratic function of three independent variables, $f(x, y, z) = 0$, are given due attention. In particular, cylinders, cones and spheres are given due attention, as these surfaces find applications in the design of multiple mechanical objects, some with rather complex forms, namely, spur and bevel gears. One more class of objects of interest in design is regular polyhedra, the 3D counterparts of regular polygons in 2D. Modern applications of such polyhedra to the design of multiaxis accelerometers, of the utmost importance for inertial navigation, are highlighted.

3.1 Points, Lines, and Planes in Space

A point P is defined in three dimensions by its three Cartesian coordinates (x, y, z), and represented by its position vector $\boldsymbol{p}$:

$$\boldsymbol{p} = \begin{bmatrix} x \\ y \\ z \end{bmatrix} \tag{3.1}$$

J. Angeles and D. Pasini, *Fundamentals of Geometry Construction*, Springer Tracts
in Mechanical Engineering, https://doi.org/10.1007/978-3-030-43131-0_3

3.1.1 Planes

A plane is the *locus*, i.e., *the set, of points equidistant from two fixed points*. The resulting locus is the perpendicular bisector of the segment joining the two points. This definition is known as *demonstrative* or *constructive*.

Computer graphics and geometric modeling require a quantitative definition. We derive below the implicit equation of the plane.

The relation between the constructive definition and the analytic-geometric definition sought can be readily derived. Let P_1 and P_2 be the two points in question, their position vectors being $\boldsymbol{p}_1$ and $\boldsymbol{p}_2$, respectively. Equating the distances, or their squares for that matter, of any point P, of position vector $\boldsymbol{p}$, to P_1 and P_2, we obtain

$$||\boldsymbol{p}_1 - \boldsymbol{p}||^2 = ||\boldsymbol{p}_2 - \boldsymbol{p}||^2$$

Each side of the above equation bears striking similarities with the square of a binomial. It is left as an exercise to the reader to prove, **without introducing components**, that the sides of that equation expand as

$$||\boldsymbol{p}_1||^2 - 2\boldsymbol{p}_1^T\boldsymbol{p} + ||\boldsymbol{p}||^2 = ||\boldsymbol{p}_2||^2 - 2\boldsymbol{p}_2^T\boldsymbol{p} + ||\boldsymbol{p}||^2$$

which readily reduces to

$$(\boldsymbol{p}_2 - \boldsymbol{p}_1)^T\boldsymbol{p} + \frac{1}{2}(||\boldsymbol{p}_1||^2 - ||\boldsymbol{p}_2||^2) = 0 \quad (3.2)$$

Now, let

$$\boldsymbol{p}_2 - \boldsymbol{p}_1 \equiv \begin{bmatrix} x_2 - x_1 \\ y_2 - y_1 \\ z_2 - z_1 \end{bmatrix} = \begin{bmatrix} A \\ B \\ C \end{bmatrix}, \quad D = \frac{1}{2}(||\boldsymbol{p}_2||^2 - ||\boldsymbol{p}_1||^2) \quad (3.3a)$$

and hence the implicit equation sought becomes

$$Ax + By + Cz + D = 0 \quad (3.3b)$$

which is a linear equation in x, y, and z. If the coordinates of any point P satisfy Eq. (3.3b), then the point is equidistant from P_1 and P_2, and hence, belongs to the plane in question. Moreover, the left-hand side of the equation being linear in the Cartesian coordinates of P, the equation defines a plane. Any point whose coordinates satisfy Eq. (3.3b) thus lies in the plane of interest.

Notice the similarity between the implicit equation of a line in 2D, Eq. (2.4), and that of a plane, as derived above. Both are *linear* in the coordinates of an arbitrary point. More importantly, Eq. (2.3) is to the plane what Eq. (3.3b) is to the three-dimensional space.

3.1.2 Lines in Space

In three-dimensional space, a line is defined by a base point A, of position vector $\boldsymbol{a}$, and a direction vector $\boldsymbol{e}$, which gives the direction of the line. Therefore, the vector equation of a line is

$$\boldsymbol{p} = \boldsymbol{a} + u\boldsymbol{e} \tag{3.4}$$

where $\boldsymbol{p}$ is the position vector of an arbitrary point P of the line and u is a real parameter. In this representation, the *sense* of the unit vector $\mathbf{e}$, given the signs of its components, is irrelevant: any change in these signs entails the corresponding change in sign of the real parameter. Unless otherwise stated, the direction vector is assumed to be of unit magnitude.

The line may also be represented in the form of three linear parametric, scalar equations, one for each coordinate:

$$\begin{aligned} x &= a_x + e_x u \\ y &= a_y + e_y u \\ z &= a_z + e_z u \end{aligned} \tag{3.5}$$

where x, y, z are the coordinates of an arbitrary point of the line, or the components of vector $\boldsymbol{p}$; hence, these coordinates are the dependent variables. Moreover, e_x, e_y, e_z are the components of the *unit vector*[1] $\mathbf{e}$; this triplet is known as the *direction cosines* of the line in question. Likewise, a_x, a_y, a_z are the components of vector $\mathbf{a}$. The set of equations in Eq. (3.5) generates a set of coordinates for each value of the parameter u. The coefficients a_x, a_y, a_z, e_x, e_y, e_z are unique and constant for any given line, modulo a sign-reversal of *all* three components of vector $\boldsymbol{e}$.

Alternatively, a line can be represented as the intersection of two planes. Each plane equation takes the form (3.3b), and hence, the two planes are represented by[2]

$$A_1 x + B_1 y + C_1 z + D_1 = 0 \tag{3.6a}$$

$$A_2 x + B_2 y + C_2 z + D_2 = 0 \tag{3.6b}$$

One point of the given line can be found upon specifying one of its three coordinates, the remaining two being found upon solving the system of Eqs. (3.6a and 3.6b) for those coordinates. Needless to say, special cases involve lines that are parallel to at least one of the three coordinate planes, X–Y, Y–Z, or X–Z. It is left as an exercise for the reader to figure out how to proceed in these cases.

---

[1] See Eq. (1.19) for a formal definition of the concept.

[2] It goes without saying that a key assumption here is that *the two planes are not parallel*.

3.1.3 Distance of a Point to a Plane

Given the plane Π represented by Eq. (3.3a) or, alternatively, by Eq. (3.3b), we want to compute the distance of a *given* point $Q(\xi, \eta, \zeta)$ to the plane, as illustrated in Fig. 3.1. To do this, we proceed as in Sect. 2.2.1: We first find the unit normal $\boldsymbol{n}$ to the plane. Moreover, let $\boldsymbol{p}_0$, $\boldsymbol{q}$, and $\boldsymbol{p}$ be the position vectors of P_0, Q, and $P(x, y, z)$, the latter being an arbitrary point of the plane. Since P_0 and P lie in the plane, the difference $\boldsymbol{p} - \boldsymbol{p}_0$ is perpendicular to the unit normal to the plane, $\boldsymbol{n}$, i.e.,

$$\boldsymbol{n}^T(\boldsymbol{p} - \boldsymbol{p}_0) = 0 \tag{3.7}$$

or, in expanded form,

$$\boldsymbol{n}^T\boldsymbol{p} - \boldsymbol{n}^T\boldsymbol{p}_0 = 0 \tag{3.8}$$

Now, let us divide both sides of Eq. (3.3b) by $\sqrt{A^2 + B^2 + C^2}$:

$$\frac{A}{\sqrt{A^2 + B^2 + C^2}}x + \frac{B}{\sqrt{A^2 + B^2 + C^2}}y + \frac{C}{\sqrt{A^2 + B^2 + C^2}}z + \frac{D}{\sqrt{A^2 + B^2 + C^2}} = 0 \tag{3.9}$$

Comparison of Eqs. (3.8) and (3.9) leads to

$$\boldsymbol{n} = \frac{1}{\sqrt{A^2 + B^2 + C^2}}\begin{bmatrix} A \\ B \\ C \end{bmatrix}, \quad \boldsymbol{n}^T\boldsymbol{p}_0 = -\frac{D}{\sqrt{A^2 + B^2 + C^2}} \tag{3.10}$$

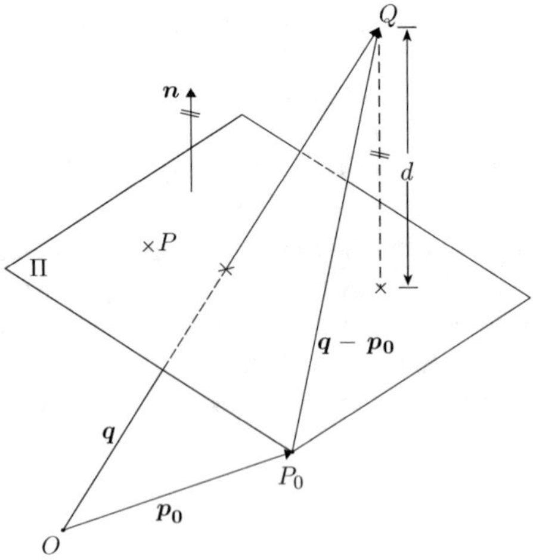

Fig. 3.1 Distance of a point to a plane

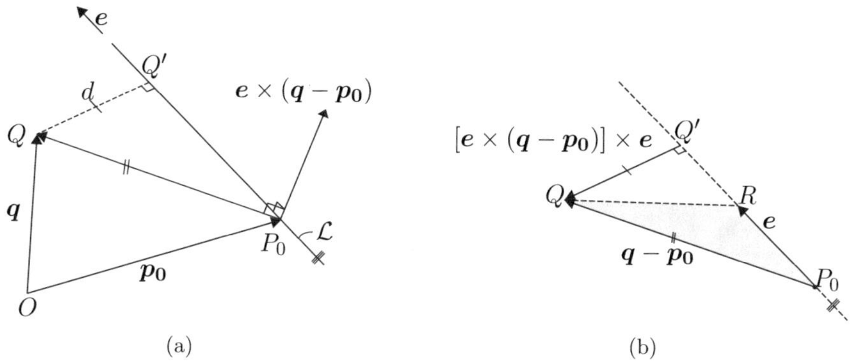

Fig. 3.2 Distance of a point to a line: **a** general layout; **b** geometric interpretation of $[\boldsymbol{e} \times (\boldsymbol{q} - \boldsymbol{p}_0)] \times \boldsymbol{e}$

From Fig. 3.1, the distance d sought is nothing but the *absolute value* of the projection of vector $\boldsymbol{q} - \boldsymbol{p}_0$ onto the unit normal $\boldsymbol{n}$, namely,

$$d = |\boldsymbol{n}^T(\boldsymbol{q} - \boldsymbol{p}_0)| \tag{3.11}$$

Interestingly, the expression for the distance of a point to a line in the 2D case, Eq. (2.9), is *formally identical* to the foregoing expression.

3.1.4 Distance of a Point to a Line

A line $\mathcal{L}$ is given by the two planes (3.6a and 3.6b). We want to find the distance of an arbitrary point $Q(\xi, \eta, \zeta)$ to $\mathcal{L}$.

First, we need a unit vector $\boldsymbol{e}$ parallel to $\mathcal{L}$ and a point P_0 of $\mathcal{L}$. If we denote by $\boldsymbol{n}_1$ and $\boldsymbol{n}_2$ the unit normal to each of the two planes that define $\mathcal{L}$, then we can obtain $\boldsymbol{e}$ as $\boldsymbol{n}_1 \times \boldsymbol{n}_2/\|\boldsymbol{n}_1 \times \boldsymbol{n}_2\|$. Moreover, $\boldsymbol{n}_1$ and $\boldsymbol{n}_2$ are produced using the expression for $\boldsymbol{n}$ displayed in Eq. (3.10):

$$\boldsymbol{n}_i = \frac{1}{\sqrt{A_i^2 + B_i^2 + C_i^2}} \begin{bmatrix} A_i \\ B_i \\ C_i \end{bmatrix}, \quad i = 1, 2 \tag{3.12}$$

We thus have the layout of Fig. 3.2a.

From Fig. 3.2a, P_0 is a point of $\mathcal{L}$, of position vector $\boldsymbol{p}_0$, while $\boldsymbol{e} \times (\boldsymbol{q} - \boldsymbol{p}_0)$ is a vector normal to the plane defined by $\mathcal{L}$ and Q, its norm $\|\boldsymbol{e} \times (\boldsymbol{q} - \boldsymbol{p}_0)\|$ being twice the area of the triangle P_0RQ depicted in Fig. 3.2b. Moreover, if we regard P_0R as the base of the triangle, d becomes its height, and hence

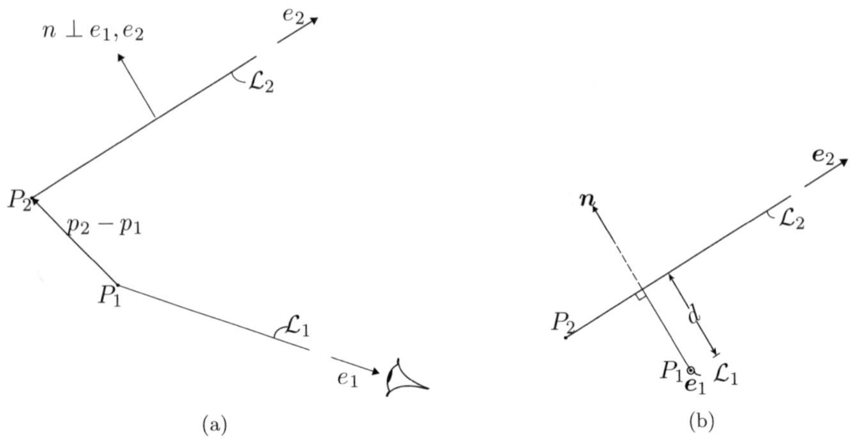

Fig. 3.3 Distance between two lines: **a** general layout; **b** view with $\mathcal{L}_1$ projected as a point, and $\boldsymbol{n}$ lying in the projection plane

$$\overline{P_0R}d = \|\boldsymbol{e} \times (\boldsymbol{q} - \boldsymbol{p}_0)\| \tag{3.13}$$

where $\overline{P_0R} = \|\boldsymbol{e}\| = 1$, so that

$$d = \|\boldsymbol{e} \times (\boldsymbol{q} - \boldsymbol{p}_0)\| \tag{3.14}$$

thereby computing the desired distance. Notice, moreover, that

$$\boldsymbol{q} - \boldsymbol{q}' = [\boldsymbol{e} \times (\boldsymbol{q} - \boldsymbol{p}_0)] \times \boldsymbol{e} \tag{3.15}$$

In the above equation, $\boldsymbol{q}'$ is the position vector of Q', the *orthogonal* projection of Q onto $\mathcal{L}$.

3.1.5 Distance Between Two Skew Lines

We have the general layout of Fig. 3.3, depicting two *skew* lines $\mathcal{L}_1$ and $\mathcal{L}_2$, parallel to the unit vectors $\boldsymbol{e}_1$ and $\boldsymbol{e}_2$ and passing through P_1 and P_2, respectively. If $\boldsymbol{n}$ denotes the unit normal to $\mathcal{L}_1$ and $\mathcal{L}_2$, then, apparently, the distance d between the two lines is nothing but the absolute value of the projection of $\boldsymbol{p}_2 - \boldsymbol{p}_1$ onto $\boldsymbol{n}$, i.e.,

$$d = |\boldsymbol{n}^T(\boldsymbol{p}_2 - \boldsymbol{p}_1)|, \quad \boldsymbol{n} \equiv \frac{\boldsymbol{e}_1 \times \boldsymbol{e}_2}{||\boldsymbol{e}_1 \times \boldsymbol{e}_2||} \tag{3.16}$$

The above relations are illustrated in Fig. 3.3b.

3.2 Surfaces

A surface is a two-dimensional set of points, extending in two directions that change from point to point, but has no thickness. We will study various types of surfaces, as described below.

The *plane*, introduced and defined in Sect. 3.1.1, is the simplest surface. That is, the plane is the *perpendicular bisector* of the *segment* defined by two given points. A plane can also be visualized as a set of lines passing through a given point and perpendicular to one given direction. In *computer graphics*, *solid objects* can be bounded by planes, forming *facets* of the solid, each facet being a polygon. In this case, the solid turns out to be a *polyhedron*. As a matter of fact, arbitrary surfaces, like airplane fuselages, are sometimes approximated, for certain computations pertaining to the solids that they enclose—volume, centroid location, etc.—by polyhedra. These polyhedra are the *finite elements* introduced to determine the stress and strain distribution on the fuselage of aircraft.[3]

In increasing order of *complexity*, the next surface is the *quadric*, namely, a surface defined, in a certain coordinate frame, by a *quadratic tri-variate polynomial*, namely,

$$F(x, y, z) = A_{11}x^2 + 2A_{12}xy + 2A_{13}xz + A_{22}y^2 + 2A_{23}yz + A_{33}z^2 + B_1x + B_2y + B_3z + C = 0 \tag{3.17}$$

which can be cast in the compact form

$$F(\boldsymbol{p}) = \boldsymbol{p}^T\boldsymbol{A}\boldsymbol{p} + \boldsymbol{b}^T\boldsymbol{p} + C = 0 \tag{3.18a}$$

where

$$\boldsymbol{A} \equiv \begin{bmatrix} A_{11} & A_{12} & A_{13} \\ A_{12} & A_{22} & A_{23} \\ A_{13} & A_{23} & A_{33} \end{bmatrix}, \quad \boldsymbol{b} \equiv \begin{bmatrix} B_1 \\ B_2 \\ B_3 \end{bmatrix} \tag{3.18b}$$

and $\boldsymbol{p}$ defined as in Eq. (3.1), with $\boldsymbol{A}$ being *symmetric*. The above expression can be cast in *homogeneous* array form, similar to Eq. (2.16), if we introduce homogeneous coordinates, namely,

$$F(p) = \boldsymbol{p}^T\boldsymbol{R}\boldsymbol{p} = 0 \tag{3.19a}$$

where

$$\boldsymbol{p} = \begin{bmatrix} x \\ y \\ z \\ 1 \end{bmatrix}, \quad \boldsymbol{R} = \begin{bmatrix} \boldsymbol{A} & \boldsymbol{b} \\ \boldsymbol{b}^T & C \end{bmatrix} \tag{3.19b}$$

[3]In fact, some "advanced" finite elements are not polyhedra, but solids bounded by combinations of quadratic or higher order simple surfaces.

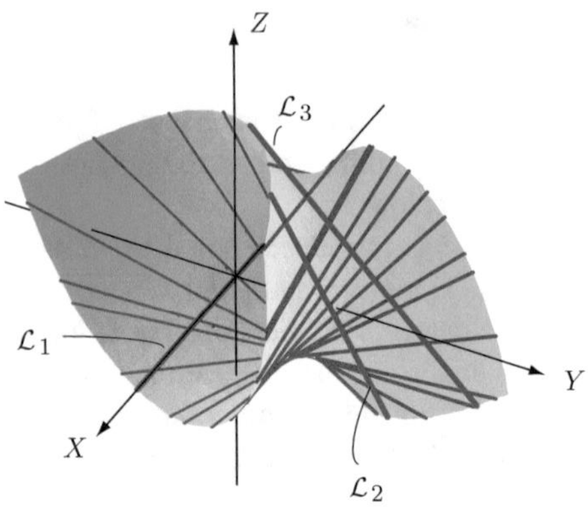

Fig. 3.4 A single-sheet hyperboloid created using three skew lines

Examples of quadrics are the *ellipsoid*, a particular case of which is the *sphere*; the *two-sheet hyperboloid*; the *single-sheet hyperboloid*; the *hyperbolic paraboloid*; and the *paraboloid*. In the same way that we have criteria based on the entries of the upper left block of matrix $\boldsymbol{R}$ of Eq. (2.17) to characterize the conics, there are criteria to characterize the quadric at hand. The reader should have noticed that matrix $\boldsymbol{R}$ of Eq. (3.19b) is *symmetric*, just like its counterpart in Eq. (2.16) in the planar case, in connection with the conic curves. The abovementioned criteria are based on matrix $\boldsymbol{A}$ of Eq. (3.19b). However, these criteria fall outside the scope of this book and will not be pursued here. The total number of quadrics[4] is 17.

Out of the foregoing surfaces, the single-sheet hyperboloid deserves special attention, as it leads, as special cases, to well-known familiar surfaces such as the cylinder and the cone. A single-sheet hyperboloid can be generated by the motion of a line constrained to intersect three given *skew lines*. Two skew lines do not intersect, which means that they do not intersect at all, not even at infinity, as do parallel lines. Shown in Fig. 3.4 is a picture of a single-sheet hyperboloid, as defined by three skew lines, $\{\mathcal{L}_i\}_1^3$.

All surfaces generated by the motion of a line belong to the class of *ruled surfaces*. The moving line generating the surface is termed the *generatrix*, which moves along a curve termed the *directrix*, the relative orientation of the generatrix with respect to the directrix being, in general, variable. Depending on the pattern of this variation, different surfaces can be obtained from the same directrix. In particular, when the orientation of the generatrix is kept constant, a *cylindrical surface* is obtained. Obviously, if the directrix is a circle and the generatrix remains normal to the plane of the circle, then a *circular cylinder* is generated. If the directrix is still a circle, but the generatrix is constrained to pass through a given point, then a *cone* is obtained. Moreover, if the given point lies on the normal to the plane of the circle passing through the center of the circle, then a *right circular cone* is generated.

Ruled surfaces are *curved* in one direction, that of the directrix, but are straight in the direction of the generatrix. For this reason, such surfaces are sometimes termed

[4]For the whole list, visit http://mathworld.wolfram.com/QuadraticSurface.html.

single-curve surfaces. Probably the most photographed ruled surface is the *Guggenheim Museum* in Bilbao, Spain. The architect behind this design, Frank Gehry, became well known in architects' circles because of his pioneering use of the CAD software package that was then in use for the design of aircraft fuselage. The design of the latter is much more daring in comparison, because the fuselage is a *double-curved* surface. The most general surfaces are *double curve*. The simplest examples of these are the ellipsoid and the paraboloid.

In computer graphics, certain special kinds of surfaces can be obtained from a generatrix. For example, to generate an axially symmetric surface, like the sphere, we can turn a circle around one of its diameters. Such surfaces are termed *surfaces of revolution*. Other means of generating surfaces is by *extrusion*, whereby a generatrix is translated in one fixed direction. Extrusion is studied in a more general form in Chap. 4.

A doubly curved surface is sometimes referred to as a *warped surface*. The best known example of such a surface is a horse saddle. Other occurrences are the human body, for example, and many engineering works like the aircraft fuselage mentioned above and the styled, aerodynamic shapes of sports cars.

3.3 Simple Solids

3.3.1 Cylinders

3.3.1.1 Description

A cylindrical surface is a ruled surface formed by a line segment called the cylinder *generatrix* that moves while remaining parallel to a fixed line, and touching a planar curve, called the cylinder *directrix*. Moreover, this curve lies in a plane intersecting the fixed line, called the cylinder axis, as shown in Fig. 3.5. The faces of the teeth of *spur* gears are cylindrical surfaces, as depicted in Fig. 3.6; these surfaces have a planar involute—see Fig. 2.19c—as their directrix. This is the reason why such gears are called *involute gears*, to distinguish them from other kinds, like *cycloidal gears*. The directrix of this alternative kind of gears is an *epicycloid*,[5] a curve illustrated in Fig. 2.19b.

If the above-mentioned generatrix remains perpendicular to the directrix, the cylinder is *straight*; otherwise, the cylinder is *oblique*. A right circular cylinder, whose directrix is a circle, is depicted in Fig. 3.5.

The representation of a *bounded* solid cylinder also calls, of course, for inequalities. For example, the solid cylinder of Fig. 3.5 is represented by

$$x^2 + z^2 \leq R^2, \quad 0 \leq y \leq h \tag{3.20}$$

[5]Only a small part of this curve is used to define the gear-tooth profile that is close to the X-axis in Fig. 2.19b.

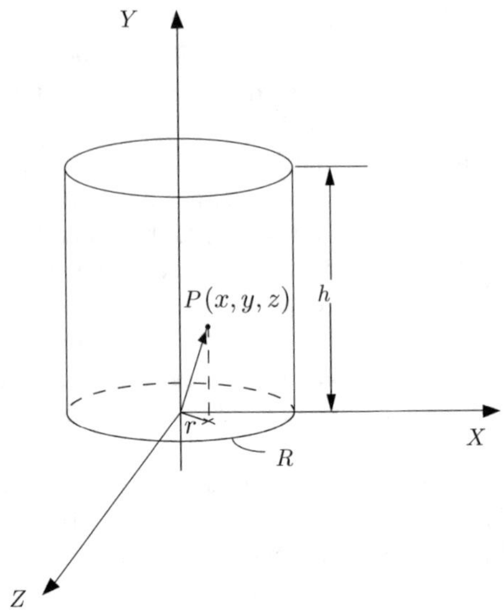

Fig. 3.5 A bounded circular cylinder

The first inequality excludes points lying a distance greater than R from the Y-axis, which is the cylinder axis; the second inequality restricts the points inside the cylinder to lie between the X–Y-plane and the plane parallel to the former a distance h above it.

In array form, the foregoing inequality becomes

$$\begin{bmatrix} x & y & z & 1 \end{bmatrix} \begin{bmatrix} 1 & 0 & 0 & 0 \\ 0 & 1 & 0 & 0 \\ 0 & 0 & 0 & 0 \\ 0 & 0 & 0 & -R^2 \end{bmatrix} \begin{bmatrix} x \\ y \\ z \\ 1 \end{bmatrix} \leq 0, \quad 0 \leq y \leq h \tag{3.21}$$

A pair of involute spur gears is used to transmit torque and motion between shafts of parallel axes. In order to design a spur-gear pair, two cylinders are produced with parallel axes and radii inversely proportional to the intended gear ratio. Next, an involute curve is produced around each of the bases of the circular cylinders. Now, with each of these curves as the directrix of a cylindrical surface, the flanks of the gears are generated. The reason behind the use of the involute lies in its geometric properties: upon meshing two involute profiles with a common tangent, each profile can turn about its cylinder axis, while maintaining a constant angle between the line of centers of the cylinder bases and the common normal. This angle is known as the *pressure angle*, which should be kept constant to obtain a constant gear ratio.

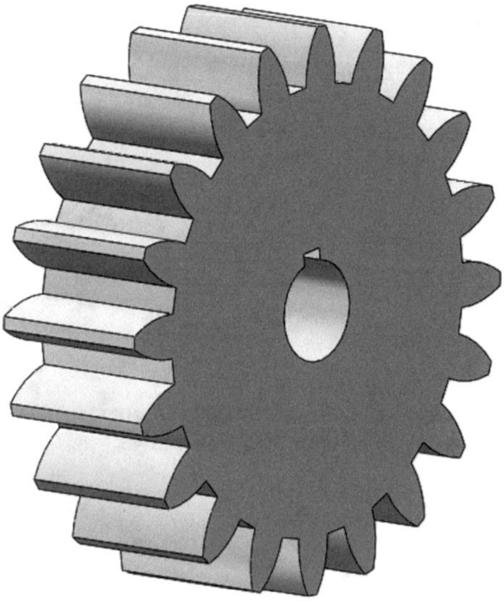

Fig. 3.6 A spur gear with its tooth flanks as instances of a cylindrical surface

3.3.2 Cones

3.3.2.1 Description

A conic surface is a ruled surface formed by a line, the cone *generatrix*, moving along a curved path, the cone *directrix*, such that the line always passes through one fixed point, called the *vertex*, a.k.a. the *apex*. Each position of the generatrix is called an *element* of the surface. Such surfaces, when bounded by a plane that cuts all elements, form *solid cones*. The surface defined by the teeth of a *bevel gear*, shown in Fig. 3.7, is a conic surface.

The simplest conic surfaces are those whose generatrix makes a *constant angle* with a fixed line, called its *axis*. Instances of solid cones produced from these surfaces are illustrated in Fig. 3.8, where we distinguish

- A *right cone*, shown in Fig. 3.8a, characterized by having its circular base normal to its axis.
- An *oblique cone*, as shown in Fig. 3.8b, whereby the axis is oblique to the circular base.
- A right cone to which a portion containing the apex has been cut off by means of a plane oblique to the axis gives rise to a *truncated cone*, as illustrated in Fig. 3.8c.
- A *frustum*, obtained if the foregoing cutting plane is normal to the axis, as depicted in Fig. 3.8d.

The flanks of the teeth of a bevel gear are made of conic surfaces. Commonly, the directrix of the conic surface in question is an *approximation* of the *spherical*

Fig. 3.7 A bevel gear with its tooth flanks as instances of conical surfaces: **a** cut with the exact spherical involute; **b** with the Tredgold approximation (Shigley and Uicker 1995) (produced by Mathew Shaker, M.Eng.)

Fig. 3.8 Various instances of solid cones: **a** right; **b** oblique; **c** truncated; **d** frustum

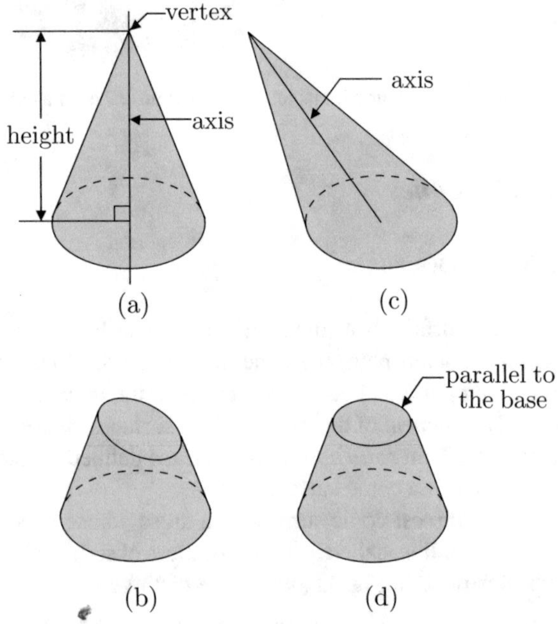

involute, namely, the spherical version of the planar involute introduced in Fig. 2.18c. In a nutshell, the planar involute is generated from a circle onto which a string has been wrapped, with one end of the string attached to one point on its circumference. Upon unwrapping the string, while maintaining it taut and lying in the plane of the circle, the free end describes a circular involute, as depicted in Fig. 2.19a. This idea can be extended to a *non-major*[6] circle lying in the surface of a sphere. Now form a right circular cone with the foregoing circle as its base, the center of the sphere as

[6]A major circle in a sphere is defined as one whose center coincides with the center of the sphere.

its apex. Further, wrap a sheet on the surface of the cone, which is possible because the cone is a single-curvature surface, and hence it is *developable*.[7] Upon fixing one straight end of the sheet to one element of the cone, then unwrapping the sheet while maintaining it taut, the sheet sweeps a second ruled surface, whose intersection with the given sphere produces a spherical curve known as the *spherical involute*.

In the times before numerically controlled (NC) machine tools, producing the conical involute surface, was a challenge. Not anymore, but traditions are hard to die. Even nowadays, the teeth of bevel gears are machined by means of the *Tredgold approximation*.[8] However, the *exact spherical involute* is making inroads in gear manufacturing.

In summary, the *pitch surfaces* of spur gears are cylinders, those of bevel gears are cones, but this is not the end of the story: spur gears are used to couple gears with parallel axes, bevel gears to couple gears with intersecting axes. Gear pairs with *skew axes* have *hyperboloids of revolution* as pitch surfaces; they are hence called *hypoid gears*. It is known that spur and bevel gear pairs can roll with respect to each other without sliding; not so hypoid gears, which both roll and slide with respect to each other. The *pitch hyperboloid*, whose parameters depend on the distance and the angle between the two gear axes and the gear ratio, is the surface that allows for *minimum sliding*, and hence *minimum power losses* due to friction.

Alas, since the advent of the automobile, which brought about the need of gears with skew axes to transmit motion from the crankshaft to the traction axle, whose axes are skew, the pitch surfaces of the two meshing gears are not hyperboloids of revolution, but, rather, cones of revolution.[9] Blame it on tradition, more so than on the state of the art of manufacturing technology!

Besides bevel gears, there are many applications for cones in engineering design, including spherical cams[10]; the nose cone of rockets; transition pieces for heating, ventilation, and air-conditioning (HVAC) systems; and conical roof sections. Cones are represented in multiview drawings by drawing the base curve, vertex, axis, and limiting elements in each view.

The representation of a cone as a bounded solid involves inequalities. For example, the right cone with apex at the origin, cone angle α and axis coincident with the Y-axis illustrated in Fig. 3.9 admits the representation:

$$x^2 + z^2 \leq k^2 y^2, \quad 0 \leq y \leq h, \quad k \equiv \tan\alpha \tag{3.22}$$

The first inequality excludes points lying at a given coordinate y outside the circle of radius $r = \sqrt{x^2 + z^2}$, which is parallel to the X–Z-plane and a distance $y \geq 0$ from this plane. The second, double, inequality restricts the points of the cone to lie in the positive region of the Y-axis *and* below or in the plane $y = h$.

---

[7]I.e., the conic surface can be laid flat on a plane.

[8]Goldfarb et al. (2018).

[9]Litvin and Fuentes (2004).

[10]See **Conic Extrusion** in Sect. 4.5.2.

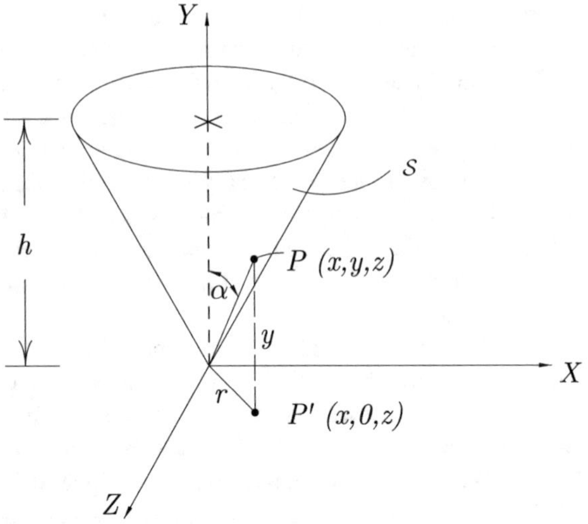

Fig. 3.9 A solid, bounded cone

The foregoing inequality is written below in array form:

$$\begin{bmatrix} x & y & z & 1 \end{bmatrix} \begin{bmatrix} 1 & 0 & 0 & 0 \\ 0 & -k^2 & 0 & 0 \\ 0 & 0 & 1 & 0 \\ 0 & 0 & 0 & 0 \end{bmatrix} \begin{bmatrix} x \\ y \\ z \\ 1 \end{bmatrix} \le 0, \quad 0 \le y \le h \tag{3.23}$$

3.3.3 Regular Polyhedra

Regular polyhedra have regular polygons for faces. There are five regular polyhedra, also known as *Platonic solids*, namely, *tetrahedron*, *hexahedron*, *octahedron*, *dodecahedron*, and *icosahedron*. Illustrated in Fig. 3.10 are the five Platonic solids.

Tetrahedron:	A solid object with four equilateral triangular facets.
Hexahedron:	A solid object with six square facets intersecting at right angles.
Octahedron:	A solid object with eight equilateral triangular facets.
Dodecahedron:	A solid object with 12 regular pentagonal facets.
Icosahedron:	A solid object with 20 equilateral triangular facets.

Just as regular polygons, the Platonic solids exhibit interesting properties that make them attractive for applications. Indeed, the centroid of every Platonic solid is located at the center of its circumscribing sphere—all vertices of such a solid are equidistant from an internal point, and hence this point is the center of the circumscribing sphere. As well, if models of the Platonic solids are constructed of a material of uniform density ρ, then their moments of inertia, 3×3 symmetric, positive-definite matrices,[11] are proportional to ρ and to the 3×3 identity matrix. By virtue of

[11] A $n \times n$ matrix $\mathbf{M}$ is said to be positive-definite if, for any n-dimensional vector, $\boldsymbol{x}$, $\boldsymbol{x}^T\mathbf{M}\boldsymbol{x} > 0$.

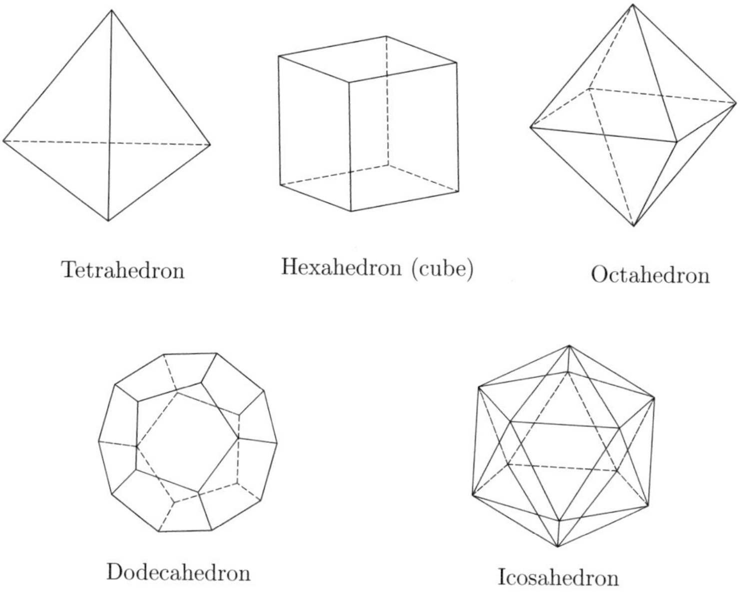

Fig. 3.10 The Platonic solids (regular polyhedra)

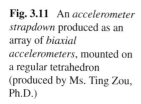

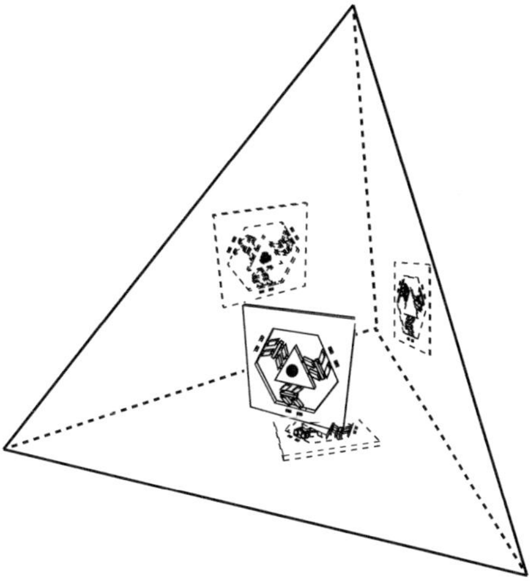

Fig. 3.11 An *accelerometer strapdown* produced as an array of *biaxial accelerometers*, mounted on a regular tetrahedron (produced by Ms. Ting Zou, Ph.D.)

this property, accelerometer strapdowns—i.e., arrays of accelerometers—have been proposed that are laid out on the faces of a Platonic solid.[12] Shown in Fig. 3.11 is a strapdown of four biaxial accelerometers like that shown in Fig. 3.12. The centroids of the accelerometers are located at the centroids of the triangular faces of a regular tetrahedron, thereby providing eight acceleration signals that are used to *estimate*—

[12]Zou and Angeles (2018).

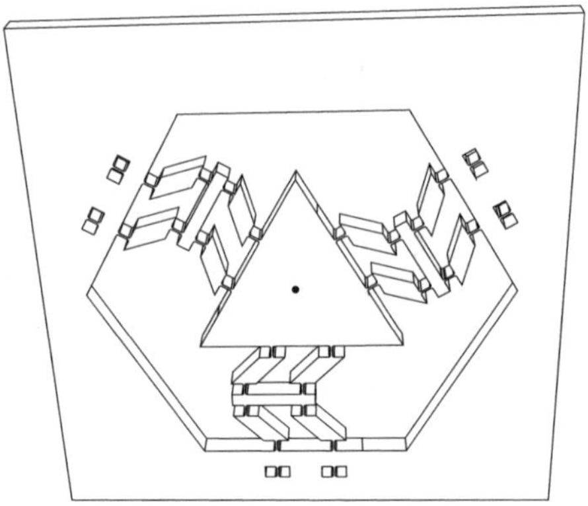

Fig. 3.12 One of the biaxial accelerometers of Fig. 3.11 (produced by Ms. Ting Zou, Ph.D.)

meaning: quantity sought inferred by means of related data picked up by sensors, complemented with a mathematical model, rather than by direct measurements—the three components of the acceleration of the tetrahedron centroid and the three of its angular acceleration. This information is then used to estimate, by time-integration, the centroid velocity and the tetrahedron angular velocity. By means of a second time-integration, the centroid position and the tetrahedron orientation are then estimated.

3.3.4 Prisms and Pyramids

3.3.4.1 Prism: Description

A polygonal prism is a polyhedron that has two equal parallel faces, called its bases, and lateral faces that are parallelograms. The parallel bases are identical closed polygons that are connected by parallelogram sides. If the parallelograms happen to be rectangles, the prism is *right*; otherwise, *oblique*. A *truncated prism* is obtained when a portion of a prism is cut off by means of a plane that intersects all the parallelogram faces, the plane being oblique to the bases. A parallelepiped is a prism with parallelograms as its bases. Polygonal prisms are readily produced with 3D CAD software by means of extrusion techniques, to be studied in Chap. 4. A few instances of prisms are shown in Fig. 3.13.

3.3.4.2 Pyramid: Description

A pyramid is a polyhedron that has a polygon for a base and lateral triangular faces that have one common intersection point, called the vertex.

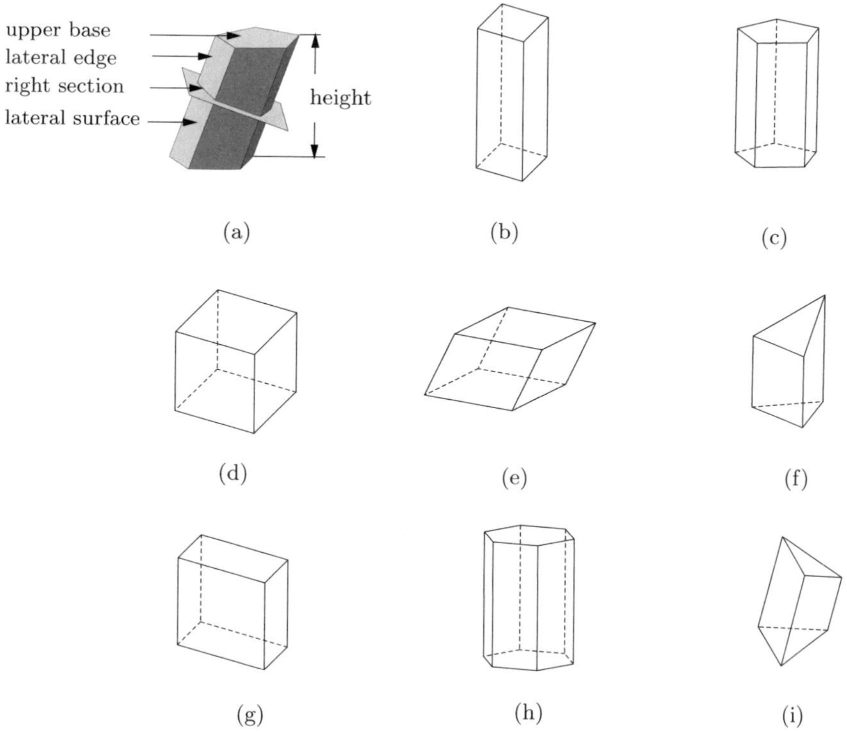

Fig. 3.13 Classification of prisms: **a** oblique pentagonal; **b** right quadrilateral; **c** right pentagonal; **d** cube; **e** oblique parallelepiped; **f** truncated; **g** right rectangular; **h** right hexagonal; **i** truncated oblique triangular

A *right pyramid* can be defined if we recall the concept of *centroid* of a planar, closed region[13]: regardless of its shape—curved, polygonal, convex, or non-convex, etc.—a contour always has a point C such that, when the contour is cut out of a sheet of material (cardboard, metal, wood) with a uniform density, and the piece of sheet is suspended from the ceiling by means of a string attached to C, the piece remains with the position and orientation at which it was released once the string is taut. Point C is the piece centroid.

If the pyramid has an arbitrary polygon as a base and the normal projection of its vertex onto its base coincides with the base centroid, then the pyramid is called *right*; otherwise, *oblique*.

If a right pyramid has a regular polygon as base, then all its sides are isosceles triangles and the pyramid is called *regular*. A *truncated pyramid* is obtained when a portion of a pyramid that contains its vertex is cut off by means of a plane at an arbitrary angle with the base.

---

[13]The concept was first invoked in Sect. 2.3.

3.4 Properties of Bounded Solids

Just as planar figures, their solid counterparts can be translated and rotated in three-dimensional space without changing their shapes and dimensions—again, as in the two-dimensional case, dimensions can be changed while preserving shape under a *uniform scaling*.[14] As discussed in Sect. 2.8 for rigid bodies in the plane, a solid rigid body in three-dimensional space is defined as a continuum set of points with one property: *under whatever displacement the body undergoes, all its points remain equidistant*. However, in principle, this property alone does not suffice to characterize the rigid body in three dimensions. A *reflection*, to be studied in Sect. 4.1.4 for the planar case and in its counterpart 4.3.4 for the three-dimensional case, a reflection also preserves distances, but reflections are not instances of a rigid-body displacement.

Similar to the two-dimensional case as well, the *pose of a rigid body* in three-dimensional space is given by (i) the position of any of its points and (ii) its *orientation*. However, unlike the two-dimensional case, in which the rigid-body orientation is defined by that of any of the body lines, in three dimensions, the body orientation is defined by *three* of its lines. To ease matters, the three lines are usually defined as the three mutually orthogonal axes of a coordinate frame with origin at one given point of the body. This frame is known as the *body frame*, to distinguish it from the *fixed coordinate frame*, also referred to in computer graphics as the *world coordinate frame*.

3.5 Composite Solids: Boolean Operations

George Boole (1815–1864) invented the algebra we use for combining sets. The Boolean operators are *union*, *intersection*, and *difference*, as illustrated in Fig. 3.14.

The **union operator** $\cup$ combines two sets, $\mathcal{A}$ and $\mathcal{B}$, to form a third set $\mathcal{C}$ whose members are either members of $\mathcal{A}$ or members of $\mathcal{B}$, or even members of both $\mathcal{A}$ *and* $\mathcal{B}$. We express this as a Boolean algebraic relation:

$$\mathcal{C} = \mathcal{A} \cup \mathcal{B} \tag{3.24}$$

For example, if $\mathcal{A} = \{a, b, c, d\}$ and if $\mathcal{B} = \{c, d, e\}$, then $\mathcal{C} = \{a, b, c, d, e\}$.

The **intersection operator** $\cap$ combines two sets $\mathcal{A}$ and $\mathcal{B}$ to form a third set $\mathcal{C}$, whose members belong to *both* $\mathcal{A}$ *and* $\mathcal{B}$, which we write as

$$\mathcal{C} = \mathcal{A} \cap \mathcal{B} \tag{3.25}$$

Using the example of sets $\mathcal{A}$ and $\mathcal{B}$, whose members were described for the union operator, we find that $\mathcal{C} = \mathcal{A} \cap \mathcal{B} = \{c, d\}$. Apparently, the intersection of two sets

[14] See Sect. 2.8.

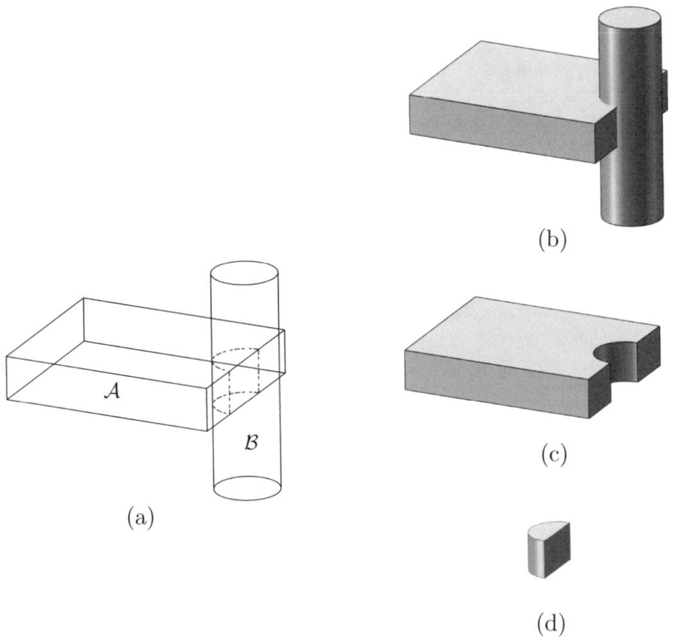

Fig. 3.14 The three Boolean operations: **a** two sets of points, $\mathcal{A}$, a rectangular parallelepiped, and $\mathcal{B}$, a right circular cylinder; **b** union; **c** difference $\mathcal{A} - \mathcal{B}$; and **d** intersection

that do not contain any common elements is *empty*. The empty set is represented as $\varnothing$. Two such sets are termed *disjoint*.

The **difference operator** combines two sets $\mathcal{A}$ and $\mathcal{B}$ to form a third set $\mathcal{C}$, whose members are only those of the first set that are not also members of the second. We write this as

$$\mathcal{C} = \mathcal{A} - \mathcal{B} \tag{3.26}$$

Again, using the example of sets $\mathcal{A}$ and $\mathcal{B}$, whose members were described previously, we find that $\mathcal{C} = \mathcal{A} - \mathcal{B} = \{a, b\}$.

The Boolean operations can be used to adjoin primitives as shown in Fig. 3.15.

The union and intersection operators are commutative, i.e.,

$$\mathcal{A} \cup \mathcal{B} = \mathcal{B} \cup \mathcal{A} \tag{3.27}$$

and

$$\mathcal{A} \cap \mathcal{B} = \mathcal{B} \cap \mathcal{A} \tag{3.28}$$

As illustrated in Fig. 3.16, the difference operator is *not* commutative, for $\mathcal{A} - \mathcal{B} = \{a, b\}$ and $\mathcal{B} - \mathcal{A} = \{e\}$. Thus,

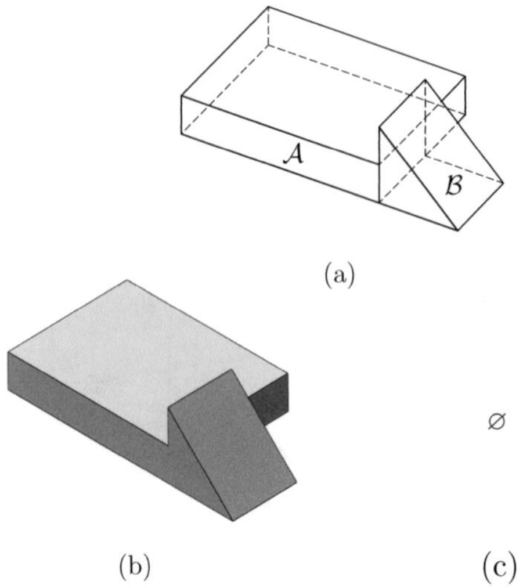

Fig. 3.15 Boolean operations on adjoining primitives: **a** two sets, $\mathcal{A}$ and $\mathcal{B}$; **b** $\mathcal{A} \cup \mathcal{B}$; **c** $\mathcal{A} \cap \mathcal{B} = \varnothing$

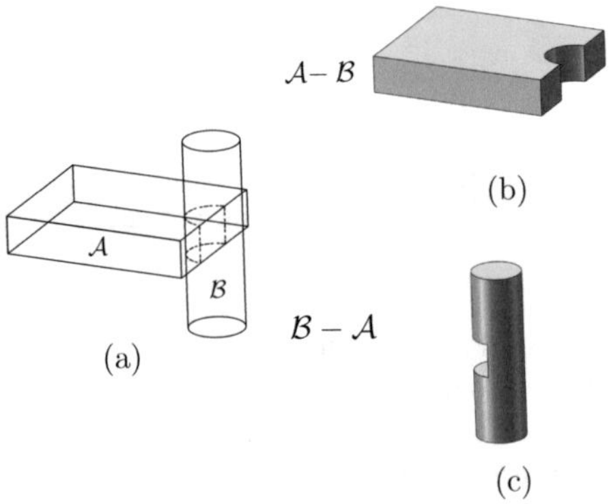

Fig. 3.16 Effects of ordering of operands in a difference operation

$$\mathcal{A} - \mathcal{B} \neq \mathcal{B} - \mathcal{A} \tag{3.29}$$

3.6 Summary

Three-dimensional objects, referred to as *rigid bodies* in mechanics, were studied in this chapter. When studying such objects, one invariably is led to study their enveloping surfaces, which are given due attention. Along these lines, the simplest surface is, of course, the plane, which is first recalled. In order of complexity, the next surface is a *quadric*, which is, as the name indicates, defined by a *quadratic* equation. Of these, the most general one is that defined by three skew lines, i.e., lines that do not share any common point—non-intersecting lines, that is. The general quadric is first introduced, which is known as the *single-sheet hyperboloid*. Its equation is systematically derived. Particular cases of these surfaces are the cylindrical and the conic surfaces. Of the former, the best known is the cylinder of circular cross section, but this is only one of infinitely many instances. Of interest to engineering are the cylindrical surfaces defining the contact surfaces of what is known as *spur gears*, which are briefly mentioned. The next type of quadric of interest to engineering is what is known as conical surfaces, of the utmost importance in the design of bevel gears. The chapter also includes commonly used surfaces in applications, such as polyhedra. The five *Platonic solids* are brought into the limelight. The chapter concludes with the concept of *Boolean operations*, of the utmost importance in the production of more general surfaces involving planar and curved surfaces.

3.7 Exercises

3.1 Using Boolean operations, describe, in set-theoretic language, how you would go about constructing the solid of Fig. 3.17, using simple elements like cylinders, prisms, or parts thereof. **N.B.**: Solution is not necessarily unique!

3.2 Repeat Exercise 3.1 for the solid of Fig. 3.18.

3.3 Derive the condition that the coefficients of Eqs. (3.6a and 3.6b) must satisfy for the two planes in question to intersect.

3.4 Three lines $\mathcal{L}_i$, $i = 1, 2, 3$ are given as[15]

(i) $\mathcal{L}_1$ is the X-axis;
(ii) $\mathcal{L}_2$ intersects the Y-axis (at right angles) a distance of $+5$ units from the origin and makes an angle of 30° with the X–Y-plane;
(iii) $\mathcal{L}_3$ passes through a point $P(1,\ 2,\ 3)$ and is parallel to the vector $\boldsymbol{v} = [-1,\ 2,\ -3]^T$.

Display the single-sheet hyperboloid defined by the three given lines. To this end, recall that the hyperboloid is traced by the motion of a fourth line $\mathcal{L}$ that is constrained to intersect the three given lines. Moreover, the distance between lines $\mathcal{L}_i$ and $\mathcal{L}$ can be computed as the *absolute value* of the projection of the difference $\boldsymbol{p} - \boldsymbol{p}_i$ onto the common normal of the two lines, parallel to $\boldsymbol{e} \times \boldsymbol{e}_i$, where

- $\boldsymbol{e}$ and $\boldsymbol{e}_i$ are unit vectors parallel to lines $\mathcal{L}$ and $\mathcal{L}_i$, respectively;
- $\boldsymbol{p}_i$ is the position vector of a point P_i of $\mathcal{L}_i$; and
- $\boldsymbol{p}$ is the position vector of a point P of $\mathcal{L}$.

Hence, the desired distance is given by

$$d = \frac{\left|(\boldsymbol{e} \times \boldsymbol{e}_i)^T(\boldsymbol{p} - \boldsymbol{p}_i)\right|}{\|\boldsymbol{e} \times \boldsymbol{e}_i\|}$$

[15] **This is an extremely challenging exercise!**

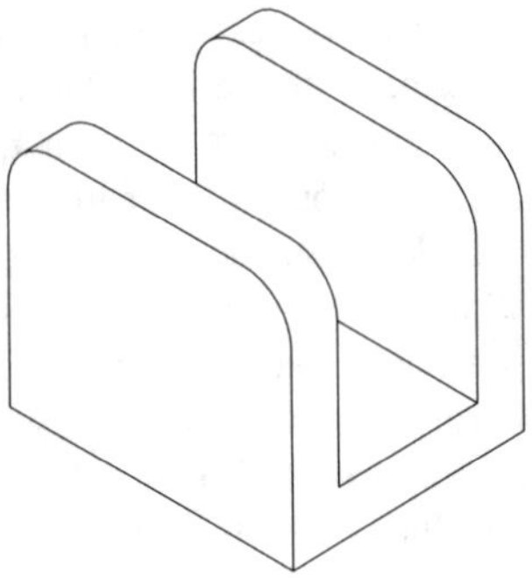

Fig. 3.17 A 3D model

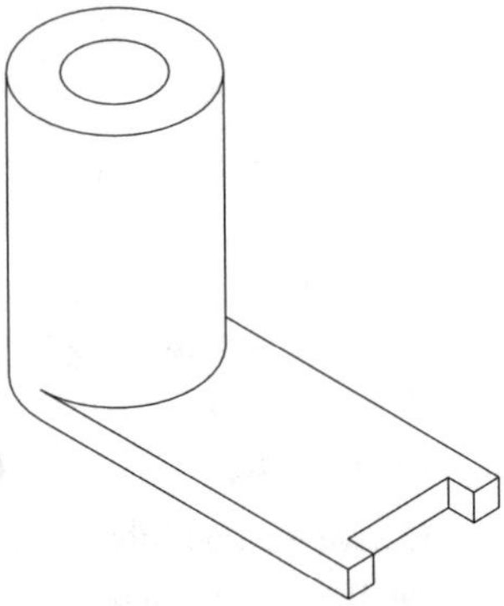

Fig. 3.18 Another 3D model

Therefore, the condition for line $\mathcal{L}$ to intersect line $\mathcal{L}_i$ can be represented as

$$(\boldsymbol{e} \times \boldsymbol{e}_i)^T(\boldsymbol{p} - \boldsymbol{p}_i) = 0, \quad i = 1, 2, 3$$

Based on the *cyclic permutability* of the above mixed product of the three vectors involved—this product was introduced in Sect. 1.4.6—we can express the above condition in the alternative form:

$$[\boldsymbol{e}_i \times (\boldsymbol{p} - \boldsymbol{p}_i)]^T \boldsymbol{e} = 0, \quad i = 1, 2, 3$$

(a) Next, we assemble the three intersection conditions displayed above to cast them into a compact form, namely,

$$\boldsymbol{A}\boldsymbol{e} = \boldsymbol{0}$$

where $\boldsymbol{A}$ is a 3×3 matrix. Display this matrix for the given data, using scientific software.

(b) Given that $\boldsymbol{e}$ is of unit magnitude, by assumption, the foregoing equation, representing a system of three *homogeneous* linear equations in three unknowns—the components of $\boldsymbol{e}$—must admit *non-trivial* solutions. Hence, matrix $\boldsymbol{A}$ must be singular, i.e.,

Fig. 3.19 A 3D model

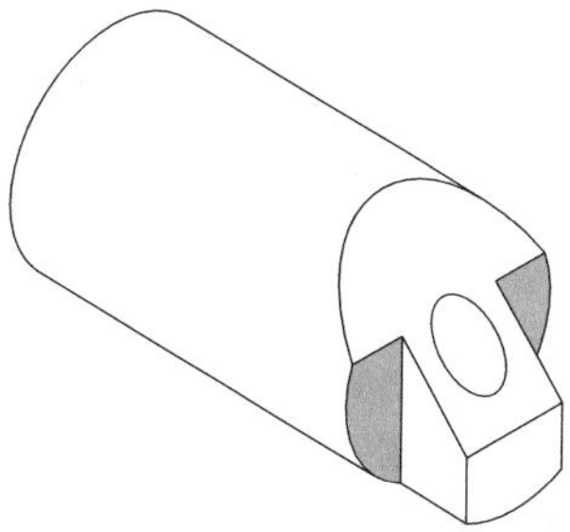

$$f(\boldsymbol{p}) \equiv \det(\boldsymbol{A}) = 0$$

which is nothing but the *implicit equation* of the hyperboloid sought. Display this equation in terms of x, y, and z, the components of $\boldsymbol{p}$, and show that it takes the general expression of a conic section, as introduced in Sect. 2.4. **Do not attempt this computation by hand!**

(c) Use the above implicit equation to display the hyperboloid given by the three lines above using scientific software. Your display should look very much like Fig. 3.4.

3.5 Using Boolean operations, describe, in set-theoretic language, how you would go about constructing the solid of Fig. 3.19, under the assumptions that: (i) the normal of the inclined plane makes an angle of 60° with the axis of the cylindrical portion; (ii) the shaded edge intersects the axis of the cylinder; and (iii) the circular cylindrical hole traverses the cylinder. **N.B.1**: Use as few simple elements as possible. **N.B.2**: Solution is not necessarily unique!

3.6 Consider the three lines $\mathcal{L}_1$, $\mathcal{L}_2$, and $\mathcal{L}_3$ passing through points Q_1, Q_2, and Q_3, of position vectors $\boldsymbol{q}_1$, $\boldsymbol{q}_2$, and $\boldsymbol{q}_3$, and having directions $\boldsymbol{v}_1$, $\boldsymbol{v}_2$, and $\boldsymbol{v}_3$, respectively, where

$$\boldsymbol{q}_1 = \begin{bmatrix} 3 \\ 2 \\ 1 \end{bmatrix}, \quad \boldsymbol{q}_2 = \begin{bmatrix} 1 \\ 3 \\ 2 \end{bmatrix}, \quad \boldsymbol{q}_3 = \begin{bmatrix} 2 \\ 1 \\ 3 \end{bmatrix},$$

$$\boldsymbol{v}_1 = \begin{bmatrix} -1 \\ 2 \\ 1 \end{bmatrix}, \quad \boldsymbol{v}_2 = \begin{bmatrix} 3 \\ -2 \\ -1 \end{bmatrix}, \quad \text{and} \quad \boldsymbol{v}_3 = \begin{bmatrix} 1 \\ -3 \\ 2 \end{bmatrix}$$

Find a set of algebraic relations among the coordinates of $P\ (x, y, z)$, allowing for the determination of point P located at a distance of 5 units from $\mathcal{L}_1$, 3 units from $\mathcal{L}_2$, and 4 units from $\mathcal{L}_3$.

N.B.: You needn't solve for the unknown coordinates, but you can do it using computer algebra.

3.7 Illustrate how to produce, with scientific software,

(a) a 2D and a 3D vector;
(b) a square matrix and a rectangular matrix;

(c) the scalar product of any two vectors, as well as the vector product of two 3D vectors; and the mixed product (a.k.a. "triple product") of three 3D vectors;
(d) the determinant of a square matrix;
(e) a 3D surface plot using the "plot" and the "implicitplot" commands.
Hint: Sometimes in computer-algebra software it is assumed that all symbolic entries of vectors in a dot product are complex, thus returning a complex product. To avoid this inconvenience, compute the dot product as the product of a row matrix by a column matrix.

3.8 (i) Two lines $\mathcal{L}_1$ and $\mathcal{L}_2$ are defined as

$$\boldsymbol{p}_1 = \boldsymbol{a}_1 + u\boldsymbol{b}_1$$

$$\boldsymbol{p}_2 = \boldsymbol{a}_2 + v\boldsymbol{b}_2$$

with

$$\boldsymbol{a}_1 = \begin{bmatrix} 6 \\ 2 \\ -3 \end{bmatrix}, \quad \boldsymbol{b}_1 = \begin{bmatrix} -1 \\ 0 \\ 2 \end{bmatrix}, \quad \text{and} \quad \boldsymbol{a}_2 = \begin{bmatrix} 3 \\ -1 \\ 2 \end{bmatrix}, \quad \boldsymbol{b}_2 = \begin{bmatrix} 3 \\ 9 \\ -3 \end{bmatrix}$$

Show that the two lines intersect.
(ii) Calculate the distance between the intersection of $\mathcal{L}_1$ and $\mathcal{L}_2$ found in (i) and the plane $\mathcal{P}$ defined as

$$3x + 2y + z - 5 = 0$$

3.9 Using Boolean operations, describe, in set-theoretic language, how you would go about constructing the solid of Fig. 3.20. The bore is a "through-hole," i.e., it traverses the cylinder.

3.10 Using a combination of (i) *surface generation by revolution*; (ii) scaling; and (iii) Boolean operations state the steps required to produce a journal bearing that will receive the conic end of a vertical shaft. The bearing has the shape of a hollow cylinder with outside diameter of 400 mm and a height of 200 mm. The conic end of the shaft will rotate on a bore with the shape of a frustum, of top diameter of 200 mm and bottom diameter of 100 mm. Frustum and cylinder are coaxial.

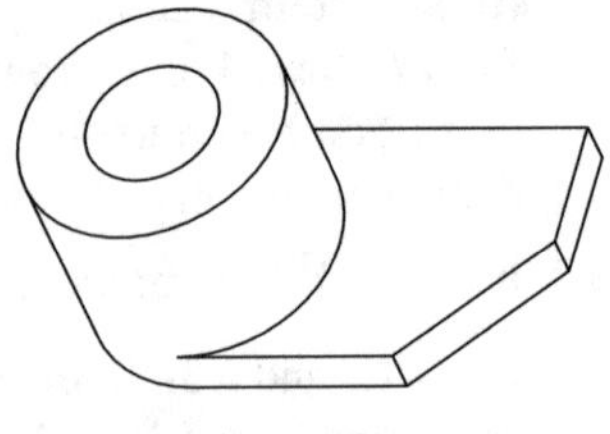

Fig. 3.20 A 3D object

References

Goldfarb V, Trubachev E, Barmina N (eds) (2018) Advanced gear engineering. Springer, New York

Litvin FL, Fuentes A (2004) Gear geometry and applied theory. Cambridge University Press, Cambridge, UK

Shigley JE, Uicker JJ Jr (1995) Theory of machines and mechanisms, 2nd edn. McGraw Hill Inc, New York

Zou T, Angeles J (2018) An algorithm for rigid-body angular velocity and attitude estimation based on isotropic accelerometer strapdowns. ASME J Appl Mech 85(6): 061010-1–061010-10. https://doi.org/10.1115/1.4039435

Chapter 4
Affine Transformations

Affine transformations allow the production of complex shapes using much simpler shapes. For example, an ellipse (ellipsoid) with axes offset from the origin of the given coordinate frame and oriented arbitrarily with respect to the axes of this frame can be produced as an affine transformation of a circle (sphere) of unit radius centered at the origin of the given frame.

An affine transformation of the space transforms bounded figures so as to: (a) enlarge or reduce their dimensions in one or two independent directions in the plane or in up to three independent directions in 3D space; (b) distort their shapes; (c) translate them; (d) rotate them; and (e) reflect them.

> *A key feature of affine transformations is that they preserve the parallelism of lines in 2D or of lines and planes in 3D.*

Affine transformations are *linear* in that they are representable by constant matrices, but not necessarily *homogeneous*, in that they can involve *translations*. Nevertheless, affine transformations can be represented as linear homogeneous transformations if homogeneous coordinates are introduced, as shown in this chapter.

4.1 2D Transformations

Let $\boldsymbol{p}$ and $\boldsymbol{p}'$ denote the three-dimensional arrays containing the homogeneous coordinates of points P and P', respectively, in the XY-plane. An affine transformation of P into P' is given by a homogeneous transformation $\boldsymbol{T}$, namely,

$$\boldsymbol{p}' = \boldsymbol{T}\boldsymbol{p} \tag{4.1a}$$

J. Angeles and D. Pasini, *Fundamentals of Geometry Construction*, Springer Tracts in Mechanical Engineering, https://doi.org/10.1007/978-3-030-43131-0_4

with $\boldsymbol{p}$, $\boldsymbol{p}'$, and $\boldsymbol{T}$ given, in turn, by

$$\boldsymbol{p} = \begin{bmatrix} x \\ y \\ 1 \end{bmatrix}, \quad \boldsymbol{p}' = \begin{bmatrix} x' \\ y' \\ 1 \end{bmatrix}, \quad \boldsymbol{T} = \begin{bmatrix} \boldsymbol{M} & \boldsymbol{t} \\ \boldsymbol{0}^T & 1 \end{bmatrix} \tag{4.1b}$$

where $\boldsymbol{t}$ is the *translation vector*, $\boldsymbol{M}$ is a 2×2 matrix defining the type of transformation at hand, and $\boldsymbol{0}$ is the two-dimensional zero vector. In affine transformations, matrix $\boldsymbol{M}$ is *non-singular*; this matrix determines the type of transformation in question, as will be made apparent in the subsections below.

The inverse of matrix $\boldsymbol{T}$ of Eq. (4.1b) was derived in Eq. (1.121) for 4×4 homogeneous transformation matrices. The 2D version of the foregoing matrix is a 3×3 matrix with the same structure as that given in Eq. (1.87), its inverse thus having the same structure as the inverse appearing in Eq. (1.121), namely,

$$\boldsymbol{T}^{-1} = \begin{bmatrix} \boldsymbol{M}^{-1} & -\boldsymbol{M}^{-1}\boldsymbol{t} \\ \boldsymbol{0}^T & 1 \end{bmatrix} \tag{4.2}$$

Since matrix $\boldsymbol{M}$ is non-singular, $\boldsymbol{T}^{-1}$ always exists.

4.1.1 Scaling

A *scaling transformation* allows an object to change by expanding or contracting its dimensions. The simplest form of scaling takes place about the coordinate axes: scaling factors S_x and S_y along the two coordinate axes provide changes in the dimensions of the object along these axes. If larger than unity, the factor represents an expansion; if smaller, a contraction. Scaling constants are always positive.

The scaling transformation of a point $P(x, y)$ into $P'(x', y')$ can be written as

$$x' = S_x x, \quad y' = S_y y, \quad \text{with} \quad S_x > 0 \quad \text{and} \quad S_y > 0 \tag{4.3}$$

In this case, then, matrix $\boldsymbol{M}$ and vector $\boldsymbol{t}$ of Eq. (4.1b) become

$$\boldsymbol{M} = \begin{bmatrix} S_x & 0 \\ 0 & S_y \end{bmatrix}, \quad \boldsymbol{t} = \boldsymbol{0} \tag{4.4}$$

which, in light of the inequalities of Eq. (4.3), is non-singular, while matrix $\boldsymbol{T}$ becomes

$$\boldsymbol{T} = \begin{bmatrix} S_x & 0 & 0 \\ 0 & S_y & 0 \\ 0 & 0 & 1 \end{bmatrix} \tag{4.5}$$

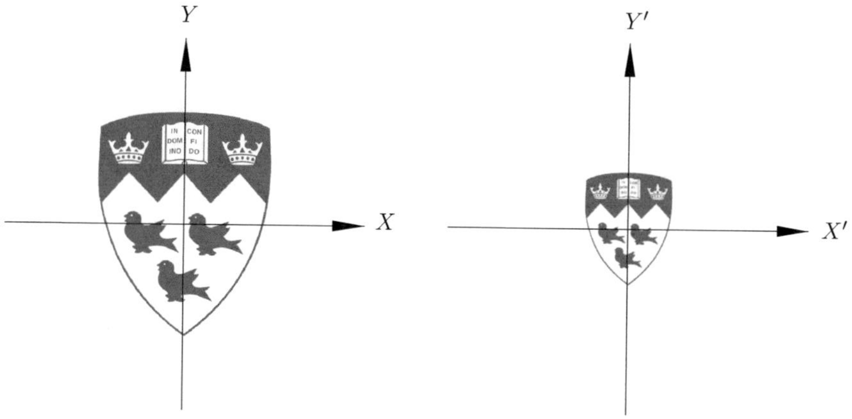

Fig. 4.1 A uniform scaling

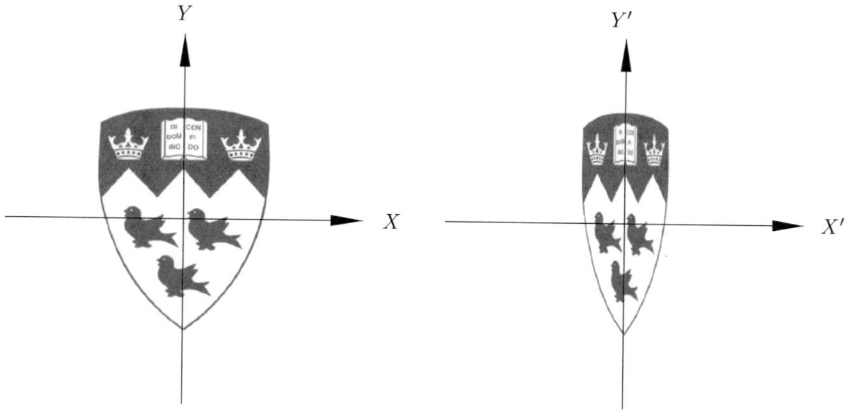

Fig. 4.2 A non-uniform scaling

The inverse of the foregoing matrices can be readily derived from the general expression (1.121), namely,

$$\boldsymbol{M}^{-1} = \begin{bmatrix} 1/S_x & 0 \\ 0 & 1/S_y \end{bmatrix}, \quad \boldsymbol{T}^{-1} = \begin{bmatrix} 1/S_x & 0 & 0 \\ 0 & 1/S_y & 0 \\ 0 & 0 & 1 \end{bmatrix} \tag{4.6}$$

Scaling is said to be *uniform*, or *isotropic*, if the scaling factors in the x- and y-directions are equal. Figure 4.1 shows an example of uniform scaling, whereas Fig. 4.2 is an example of non-uniform scaling, with a contraction in the horizontal direction.

In some instances, scaling is needed about two orthogonal axes other than the given coordinate axes. In this case, the scaling matrix $\boldsymbol{M}$ is no longer diagonal and must be calculated from the data on the desired transformation. Notice that scaling may be needed about the two orthogonal directions of a coordinate system whose origin may coincide or not with the coordinate axes. How to obtain the desired transformation matrix is the subject of Sect. 4.2.

4.1.2 Translation

The ability to move parts of a model is an essential feature of any graphics system. Translations cause an object to be displaced in a specific direction by a specific amount, while preserving its shape, size, and orientation. The translation of the point $P(x, y)$ into $P'(x', y')$ is expressed as

$$x' = x + t_x, \quad y' = y + t_y \tag{4.7}$$

In this case, matrix $\boldsymbol{M}$ and vector $\boldsymbol{t}$ of Eq. (4.1b) become

$$\boldsymbol{M} = \begin{bmatrix} 1 & 0 \\ 0 & 1 \end{bmatrix}, \quad \boldsymbol{t} = \begin{bmatrix} t_x \\ t_y \end{bmatrix} \tag{4.8}$$

in which $\boldsymbol{M}$ is the 2×2 identity matrix because of shape-, size-, and orientation-preservation. Matrix $\boldsymbol{T}$ becomes, in turn,

$$\boldsymbol{T} = \begin{bmatrix} 1 & 0 & t_x \\ 0 & 1 & t_y \\ 0 & 0 & 1 \end{bmatrix} \tag{4.9}$$

The advantage of homogeneous coordinates is apparent here: with Cartesian coordinates, rigid-body translations could not be represented in *homogenous* form.

Figure 4.3 shows an example of translation.

The inverse transformation of $\boldsymbol{T}$ given above is readily obtained from the general expression (4.2):

$$\boldsymbol{T}^{-1} = \begin{bmatrix} 1 & 0 & -t_x \\ 0 & 1 & -t_y \\ 0 & 0 & 1 \end{bmatrix} \tag{4.10}$$

Fig. 4.3 The McGill University crest undergoing a translation

4.1.3 Rotation

Rotations are frequently used to enable the viewer to see an object from different directions, or to assemble various geometric objects.

A rotation can be assumed to take place about the origin of the coordinate system by a specified angle θ. Should a rotation take place about a point other than the origin, then the corresponding transformation could be represented as a combination of translation and rotation.

Since we need a convention about the direction of rotation, we consider that counterclockwise rotations are positive, while their clockwise counterparts are negative.

We derive the rotation transformation via the polar coordinates of P, which are

$$x = r\cos\phi, \quad y = r\sin\phi \tag{4.11}$$

where ϕ is the angle and r is the radius.

The transformed position $P'(x', y')$ of point P due to the rotation can be calculated by the use of simple trigonometric relations:

$$\begin{aligned} x' &= r\cos(\phi+\theta) = r\cos\phi\cos\theta - r\sin\phi\sin\theta \\ y' &= r\sin(\phi+\theta) = r\sin\phi\cos\theta + r\cos\phi\sin\theta \end{aligned} \tag{4.12}$$

where x and y, as given by Eq. (4.11), can be readily identified. Hence,

$$x' = x\cos\theta - y\sin\theta, \quad y' = x\sin\theta + y\cos\theta \tag{4.13}$$

or, in array form,

$$\begin{bmatrix} x' \\ y' \\ 1 \end{bmatrix} = \begin{bmatrix} \cos\theta & -\sin\theta & 0 \\ \sin\theta & \cos\theta & 0 \\ 0 & 0 & 1 \end{bmatrix} \begin{bmatrix} x \\ y \\ 1 \end{bmatrix} \tag{4.14}$$

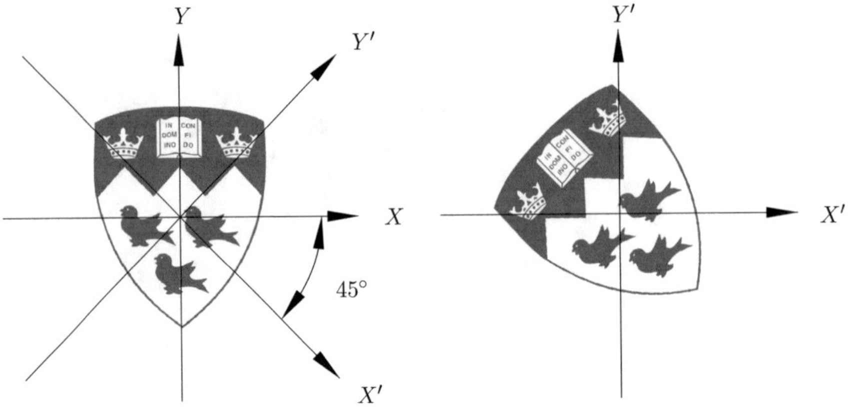

Fig. 4.4 A rotation by $\theta = 45°$

and hence, $\boldsymbol{M}$, $\boldsymbol{t}$, and $\boldsymbol{T}$ become

$$\boldsymbol{M} = \begin{bmatrix} \cos\theta & -\sin\theta \\ \sin\theta & \cos\theta \end{bmatrix}, \quad \boldsymbol{t} = \boldsymbol{0}, \quad \boldsymbol{T} = \begin{bmatrix} \cos\theta & -\sin\theta & 0 \\ \sin\theta & \cos\theta & 0 \\ 0 & 0 & 1 \end{bmatrix} \tag{4.15}$$

Figure 4.4 shows a rotation of the coordinate axes through an angle of $\theta = 45°$ cw. By the same token, the object, assumed fixed to the original axes, rotates through an angle of $\theta = 45°$ ccw.

The inverse homogeneous transformation of $\boldsymbol{T}$ of Eq. (4.15), again, can be readily derived from the general expression (4.2), while taking into account that $\boldsymbol{M}$, in this case, is orthogonal, and hence its inverse is simply its transpose, i.e.,

$$\boldsymbol{T}^{-1} = \begin{bmatrix} \cos\theta & \sin\theta & 0 \\ -\sin\theta & \cos\theta & 0 \\ 0 & 0 & 1 \end{bmatrix} \tag{4.16}$$

4.1.4 Reflection

The concept of reflection can be understood by thinking of images in a mirror. The reflection transformation is useful in the construction of symmetric objects. For example, one half of a symmetric object may be created and then conveniently reflected to generate the whole object.

In 2D, reflections are defined about a line. The reflection matrix relative to either the X- or the Y-axes can be expressed in the form of Eq. (4.1b), with $\boldsymbol{M}$ *improper orthogonal* and $\boldsymbol{t} = \boldsymbol{0}$. Improper orthogonality means that $\boldsymbol{M}$, besides being orthogonal, has a negative determinant, i.e.,

$$\boldsymbol{M}^T\boldsymbol{M} = \boldsymbol{M}\boldsymbol{M}^T = \mathbf{1}, \quad \det(\boldsymbol{M}) = -1 \tag{4.17}$$

Below we show different instances of reflections:

- About the X-axis,

$$\boldsymbol{M}_X = \begin{bmatrix} 1 & 0 \\ 0 & -1 \end{bmatrix}, \quad \boldsymbol{t} = \mathbf{0} \tag{4.18}$$

  This reflection is illustrated in Fig. 4.5.
- About the Y-axis,

$$\boldsymbol{M}_Y = \begin{bmatrix} -1 & 0 \\ 0 & 1 \end{bmatrix}, \quad \boldsymbol{t} = \mathbf{0} \tag{4.19}$$

  This reflection is illustrated in Fig. 4.6.
- A composite reflection
  As the reader might expect, the combination of the two foregoing reflections is represented by the matrix

$$\boldsymbol{M} = \begin{bmatrix} -1 & 0 \\ 0 & -1 \end{bmatrix}, \quad \boldsymbol{t} = \mathbf{0} \tag{4.20}$$

  which turns out to be a rotation about the origin through 180°, as depicted in Fig. 4.7. It should be apparent that the product of two reflections, given by improper orthogonal matrices, is necessarily a rotation, represented by a proper orthogonal matrix. Indeed, as stated—without a proof, which lies outside the scope of the book—in Sect. 1.4.6, around Eq. (1.84), the determinant of a product of square matrices is the product of the individual determinants. In this light, the product of an odd number of reflections is another reflection. However, if the number of reflections is even, then the product is a rotation.
  Other reflections through arbitrary lines are also possible. For example, the reflection about the line $y = x$ is represented by

$$\boldsymbol{M} = \begin{bmatrix} 0 & 1 \\ 1 & 0 \end{bmatrix}, \quad \boldsymbol{t} = \mathbf{0} \tag{4.21}$$

  An example of reflection about the line $y = x$ is illustrated in Fig. 4.8.

Thus, the general form of the homogeneous 3×3 transformation matrix representing a reflection does not obey a specific pattern. This matrix is exactly as displayed in Eq. (4.1b) for the general homogeneous transformation in 2D, except that $\boldsymbol{M}$ is improper orthogonal. In this light, $\boldsymbol{M}^{-1} = \boldsymbol{M}^T$, and hence

$$\boldsymbol{T}^{-1} = \begin{bmatrix} \boldsymbol{M}^T & \mathbf{0} \\ \mathbf{0}^T & 1 \end{bmatrix} \tag{4.22}$$

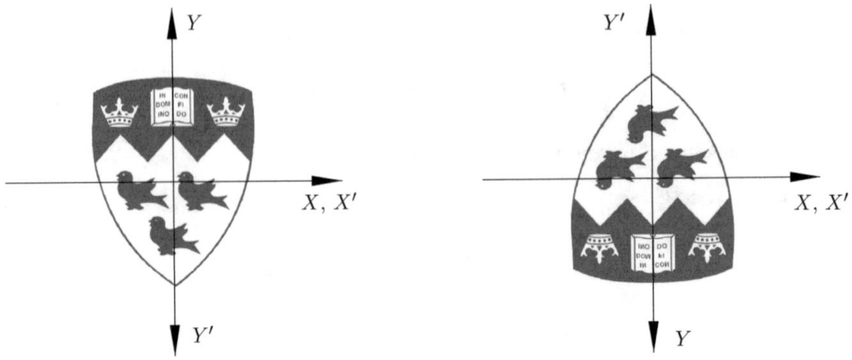

Fig. 4.5 A reflection about the X-axis

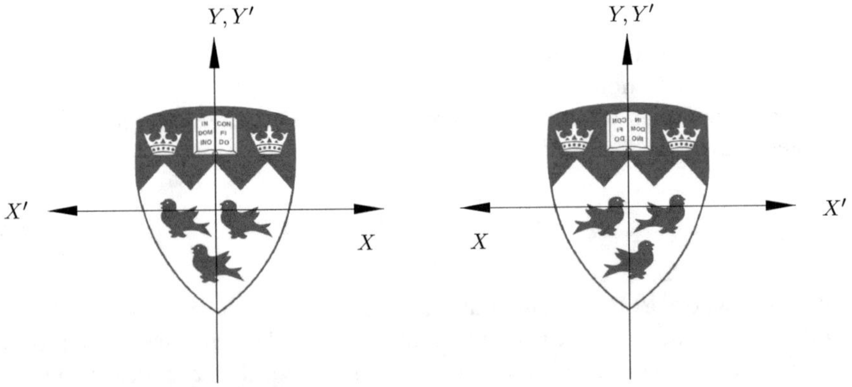

Fig. 4.6 A reflection about the Y-axis

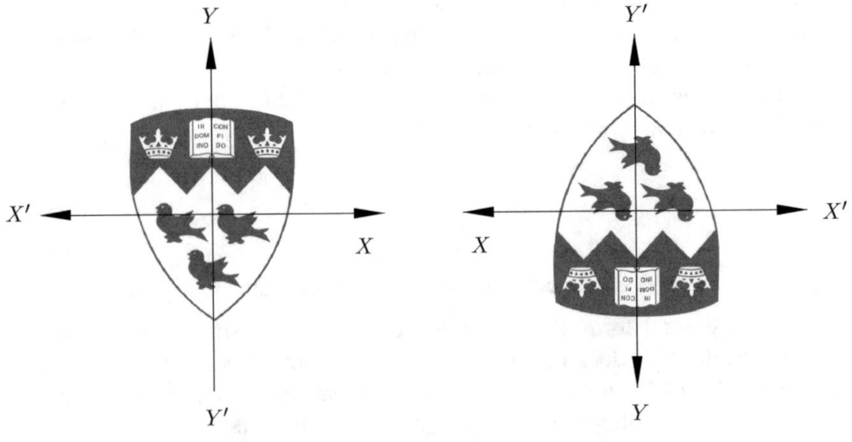

Fig. 4.7 The composition of one reflection about the X-axis with one about the Y-axis, equivalent to a rotation about the origin through 180°

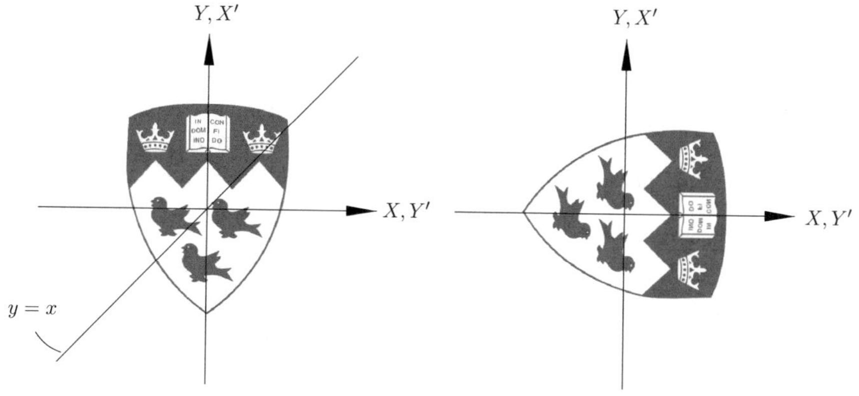

Fig. 4.8 A reflection about the line $y = x$

4.2 The Most General Affine Transformation in 2D

The most general affine transformation involves compositions of the four *simple affine transformations* studied in Sect. 4.1. The simplest way of obtaining the desired transformation is described below:

Problem 4.2.1 Given a 2D object $\mathcal{O}$, find the 3×3 homogeneous transformation matrix that transforms this object into $\mathcal{O}'$.

We distinguish here between two cases: (i) the transformation leaves the origin fixed and (ii) the transformation carries the origin to a new position, O'.

Case (i): Given that no translation of the origin is involved, the problem reduces to finding matrix $\boldsymbol{M}$ of the homogeneous transformation, and hence we have to find four unknowns, i.e., the four distinct entries m_{11}, m_{12}, m_{21}, and m_{22}. We thus need four equations relating the two objects, which can be derived from the correspondence between two points in $\mathcal{O}$ and their *images* in $\mathcal{O}'$. Let these points be P_1 and P_2 in $\mathcal{O}$ and their images, P_1' and P_2' in $\mathcal{O}'$. As each point correspondence entails two scalar equations, one for each coordinate, we have a total of four scalar equations, enough to compute the four unknowns. Let, moreover, $\boldsymbol{p}_i$ and $\boldsymbol{p}_i'$ be the position vector of P_i and P_i'. Hence,

$$\boldsymbol{p}_1' = \boldsymbol{M}\boldsymbol{p}_1, \quad \boldsymbol{p}_2' = \boldsymbol{M}\boldsymbol{p}_2 \tag{4.23a}$$

Now we introduce two 2×2 matrix arrays:

$$\boldsymbol{P} \equiv \left[\boldsymbol{p}_1 \ \boldsymbol{p}_2\right], \quad \boldsymbol{P}' \equiv \left[\boldsymbol{p}_1' \ \boldsymbol{p}_2'\right] \tag{4.23b}$$

In this way, the two vector equations (4.23a) can be cast in the form of one single 2×2 *matrix equation*:

$$P' = MP \tag{4.24}$$

from which we can solve for M by multiplying both sides by P^{-1} from the right, *provided that* P *is not singular*. This will be the case *if the two points chosen,* P_1 *and* P_2*, are not aligned with the origin*. With this provision, then, Eq. (4.24) can be solved for M in the form

$$M = P'P^{-1} \tag{4.25}$$

thereby completing the solution.

Case (ii): In this case, we have six unknowns, the four entries of M, as in Case (i), plus the two components of vector t. We thus need six scalar equations to find the foregoing unknowns. These equations are readily obtained by establishing the correspondence between two triplets of points, one in $\mathcal{O}$ and one in $\mathcal{O}'$. This correspondence should be established between homogeneous coordinates, as t is not included in M. We proceed as in Case (i): let now p_i and p_i' be the arrays of *homogeneous coordinates* of P_i and P_i', respectively, for $i = 1, 2, 3$, and T the 3×3 homogeneous-transformation matrix sought. The three correspondences then follow:

$$p_i' = Tp_i, \quad \text{for} \quad i = 1, 2, 3 \tag{4.26}$$

By defining the 3×3 arrays P and P' as

$$P = \begin{bmatrix} p_1 & p_2 & p_3 \end{bmatrix}, \quad P' = \begin{bmatrix} p_1' & p_2' & p_3' \end{bmatrix} \tag{4.27}$$

the three correspondences (4.26) can now be cast in the form of a 3×3 *matrix equation*:

$$P' = TP \tag{4.28}$$

from which we can solve for T upon multiplying both sides of the foregoing equation by P^{-1} from the right, as long, of course, as P is not singular. Notice that, although homogeneous transformations representing affine transformations are never singular, matrices P and P' can become singular. As the reader can readily verify, these matrices become singular if and only if the three given points are collinear. If this is not the case, then Eq. (4.28) can be solved for T in the form

$$T = P'P^{-1} \tag{4.29}$$

thereby completing the solution of the second case.

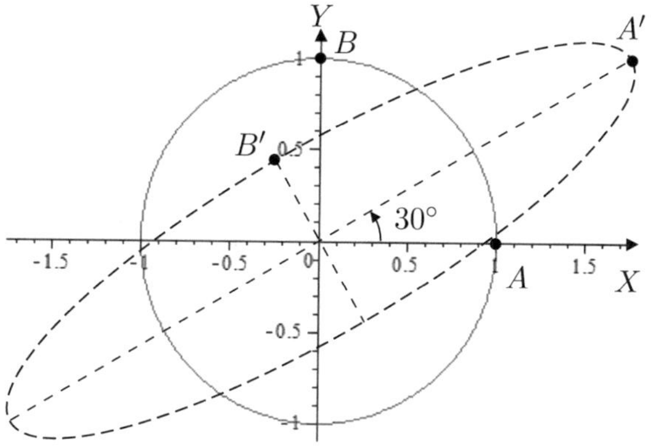

Fig. 4.9 An affine transformation of the unit circle centered at the origin into an ellipse of semiaxes 2 and 0.5, centered likewise

4.2.1 Examples

Example 4.2.1 Find the affine transformation that carries the unit circle centered at the origin of the XY-plane into the ellipse of semiaxes 0.5 and 2, centered at the origin as well, with the focal axis making an angle of 30° with the X-axis, as illustrated in Fig. 4.9. Find also the inverse transformation that carries the ellipse back into the original unit circle.

Solution We have a scaling about two orthogonal axes rotated 30° ccw from the coordinate axes, intersecting at the origin. As the scaling occurs about axes other than the given X and Y, we have to find the matrix $\boldsymbol{M}$ of the affine transformation. To this end, we define points A and B on the circle as those at which the coordinate axes intersect the circle, as indicated in Fig. 4.9. We assume that these points are *mapped* into A' and B', respectively, on the major and minor axes of the ellipse. A' (B') is said to be the *image* of A (B) under the mapping represented by $\boldsymbol{M}$. Moreover, $\boldsymbol{a}$ and $\boldsymbol{b}$ be the position vectors of A and B, respectively, the position vectors $\boldsymbol{a}'$ and $\boldsymbol{b}'$ being defined correspondingly. Thus,

$$\boldsymbol{a} = \begin{bmatrix} 1 \\ 0 \end{bmatrix}, \quad \boldsymbol{b} = \begin{bmatrix} 0 \\ 1 \end{bmatrix}, \quad \boldsymbol{a}' = \begin{bmatrix} 2\cos(\pi/6) \\ 2\sin(\pi/6) \end{bmatrix}, \quad \boldsymbol{b}' = \begin{bmatrix} 0.5\cos(2\pi/3) \\ 0.5\sin(2\pi/3) \end{bmatrix}$$

matrices $\boldsymbol{P}$ and $\boldsymbol{P}'$ being defined as

$$\boldsymbol{P} = [\boldsymbol{a}\ \boldsymbol{b}] = \begin{bmatrix} 1\ 0 \\ 0\ 1 \end{bmatrix}, \quad \boldsymbol{P}' = [\boldsymbol{a}'\ \boldsymbol{b}'] = \begin{bmatrix} 1.732 & -0.2500 \\ 1 & 0.4330 \end{bmatrix}$$

Given that $\boldsymbol{P}$ is the 2×2 identity matrix, its inverse is simply the identity itself, and hence the matrix $\boldsymbol{M}$ sought is

$$M = P'P^{-1} = \begin{bmatrix} 1.732 & -0.2500 \\ 1.000 & 0.4330 \end{bmatrix}$$

The homogeneous matrix T implementing the desired transformation is, thus,

$$T = \begin{bmatrix} 1.732 & -0.2500 & 0 \\ 1.000 & 0.4330 & 0 \\ 0 & 0 & 1 \end{bmatrix}$$

The inverse transformation of the foregoing mapping is obtained directly from Eq. (1.121), as applied to 3×3 homogeneous matrices, i.e.,

$$T^{-1} = \begin{bmatrix} M^{-1} & \mathbf{0} \\ \mathbf{0}^T & 1 \end{bmatrix}$$

where computing M^{-1} is straightforward, as this is a 2×2 matrix. With this inverse calculated, the inverse sought becomes

$$T^{-1} = \begin{bmatrix} 0.433 & 0.250 & 0 \\ -1.000 & 1.732 & 0 \\ 0 & 0 & 1 \end{bmatrix}$$

thereby completing the example.

Example 4.2.2 Find the affine transformation that carries the unit circle centered at the origin of the XY-plane into the ellipse of semiaxes 2 and 1, centered at the point $O'(3,\ 2)$, with its major axis making an angle of 60° with the X-axis, as illustrated in Fig. 4.10. Then find the inverse transformation that carries the offset ellipse back into the original unit circle.

Solution Because of the offset of the center of the ellipse, the problem involves six unknowns, and hence six equations are needed. We then proceed as in Case (ii) above.

First, the points among which the correspondence will be established are chosen. Because of the symmetry of the circle, any pair of points will do. However, if these are going to be made to correspond with points on the ellipse at the intersection with its axes, the position vectors of the points on the circle might as well be at right angles. To ease matters, we choose the same points A and B as in Example 4.2.1. The third point required in this case is obvious: the origin O of position vector $\boldsymbol{o} = \mathbf{0}$. The images of these points, A', B', and O', are taken as shown in Fig. 4.10, i.e.,

$$\boldsymbol{a} = \begin{bmatrix} 1 \\ 0 \end{bmatrix},\ \boldsymbol{b} = \begin{bmatrix} 0 \\ 1 \end{bmatrix},\ \boldsymbol{o} = \mathbf{0},\ \boldsymbol{a}' = \begin{bmatrix} 3 + 2\sin(\pi/3) \\ 2 + 2\cos(\pi/3) \end{bmatrix},\ \boldsymbol{b}' = \begin{bmatrix} 3 + \cos(5\pi/6) \\ 2 + \sin(5\pi/6) \end{bmatrix},\ \boldsymbol{o}' = \begin{bmatrix} 3 \\ 2 \end{bmatrix}$$

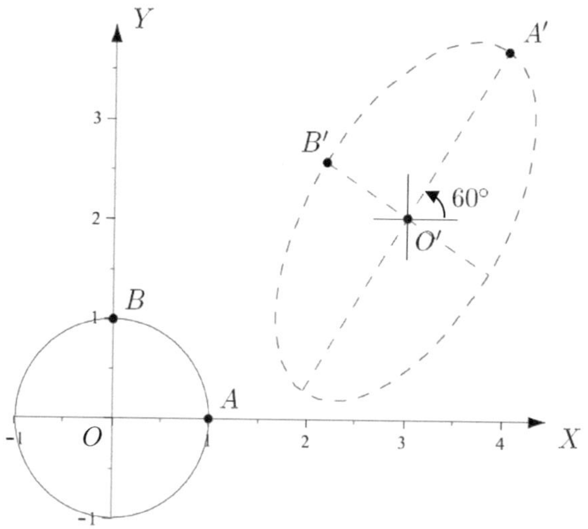

Fig. 4.10 An affine transformation of the unit circle centered at the origin into an ellipse of semiaxes 2 and 1, with its center offset from that of the circle

the 3×3 arrays of the homogeneous coordinates of the three given points, matrices $\boldsymbol{P}$ and $\boldsymbol{P}'$ being readily obtained:

$$\boldsymbol{P} = \begin{bmatrix} 1 & 0 & 0 \\ 0 & 1 & 0 \\ 1 & 1 & 1 \end{bmatrix}, \quad \boldsymbol{P}' = \begin{bmatrix} 4.000 & 2.134 & 3 \\ 3.732 & 2.500 & 2 \\ 1 & 1 & 1 \end{bmatrix}$$

Now, in order to compute matrix $\boldsymbol{T}$, we need $\boldsymbol{P}^{-1}$, where $\boldsymbol{P}$ is a 3×3 matrix with a block equal to the 2×2 identity matrix. If this matrix is written in block form as

$$\boldsymbol{P} = \begin{bmatrix} \mathbf{1} & \mathbf{0} \\ \boldsymbol{c}^T & 1 \end{bmatrix}$$

and blocks $\mathbf{1}$, $\mathbf{0}$, $\boldsymbol{c}^T$, and 1 are identified with blocks $\boldsymbol{A}$, $\boldsymbol{B}$, $\boldsymbol{C}$, and $\boldsymbol{D}$, respectively, then the formula (1.119) for the inverse of a block matrix can be readily applied, and hence

$$\boldsymbol{P}^{-1} = \begin{bmatrix} \mathbf{1} & \mathbf{0} \\ -\boldsymbol{c}^T & 1 \end{bmatrix} = \begin{bmatrix} 1 & 0 & 0 \\ 0 & 1 & 0 \\ -1 & -1 & 1 \end{bmatrix}$$

Therefore,

$$\boldsymbol{T} = \boldsymbol{P}'\boldsymbol{P}^{-1} = \begin{bmatrix} 1 & -0.8660 & 3 \\ 1.732 & 0.5000 & 2 \\ 0 & 0 & 1 \end{bmatrix}$$

its inverse being readily computed in block form using the general expression (4.2), which leads to

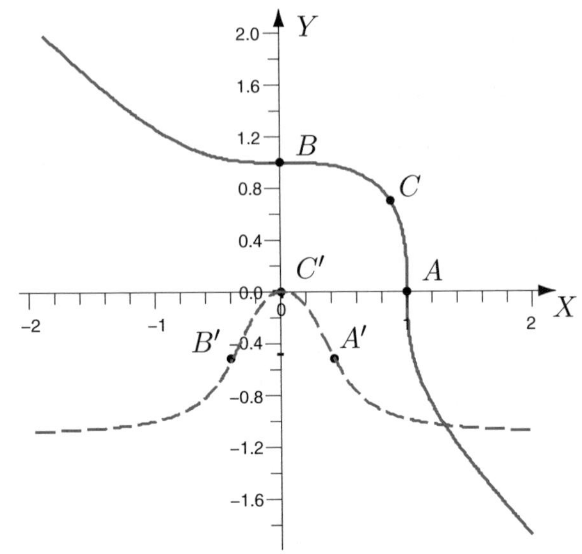

Fig. 4.11 The affine transformation of a cubic Lamé curve into a displaced, squeezed configuration

$$
\boldsymbol{T}^{-1} = \begin{bmatrix} 0.2500 & 0.4330 & -1.6160 \\ -0.8660 & 0.5000 & 1.598 \\ 0 & 0 & 1 \end{bmatrix}
$$

thereby completing the computations.

Example 4.2.3 Find the affine transformation that carries the cubic Lamé curve, displayed in Fig. 2.17b, into a configuration whereby: (a) its "bump"[1] lies at the origin of the X–Y-plane; (b) its tangent at the "bump" coincides with the X-axis; and (c) its tangents at the inflection points[2] make an angle of $2\sin^{-1}(\sqrt{5}/5) \approx 53.13°$, as displayed in Fig. 4.11. Then, find the inverse homogeneous transformation that will carry the distorted, displaced curve to its original location and shape.

Solution With reference to Fig. 4.12, the curve is transformed by (a) a distortion under which the curve is squeezed in the X''-direction, while its dimensions in the Y''-direction are preserved and (b) a displacement taking the $\{\,O'',\ X'',\ Y''\,\}$ frame to a configuration in which it coincides with the $\{\,O,\ X,\ Y\,\}$ frame. Notice, moreover, that the bump occurs at a point on the original curve at which $x = y = a$, where a is still to be found. In this light, the distance from the bump to the origin is $d = \sqrt{2}a$.

Upon setting $x = y = a$ in the equation of the cubic Lamé curve, an equation in a is found, whose real root is

[1] Faute-de-mieux, we term *bump* the point of a Lamé curve where its curvature is a maximum.

[2] These are the points at which the curvature of a *continuous* and *smooth* curve changes sign. In the case at hand, these points are A' and B'.

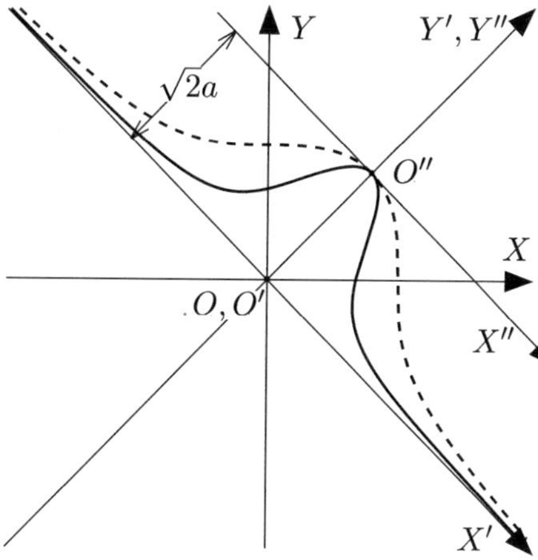

Fig. 4.12 A squeezed cubic Lamé curve

$$a = \left(\frac{1}{2}\right)^{1/3} \quad \Rightarrow \quad d = 2^{1/6} \approx 1.1225$$

Rather than obtaining the affine transformation of interest as a combination of non-uniform scaling, translation, and rotation, we use Case (ii) of the procedure outlined above. Since the transformation involves a shift of the origin—that of the $\{O'', X'', Y''\}$ frame—we need three points in the original and the transformed curves. Obvious choices are points A, B, and C of Fig. 4.11, defined as the intersections of the curve with the X- and the Y-axes plus the bump. The position vectors of these points are, thus,

$$\boldsymbol{a} = \begin{bmatrix} 1 \\ 0 \end{bmatrix}, \ \boldsymbol{b} = \begin{bmatrix} 0 \\ 1 \end{bmatrix}, \ \boldsymbol{c} = \begin{bmatrix} a \\ a \end{bmatrix} = \begin{bmatrix} 0.7937 \\ 0.7937 \end{bmatrix}, \ \boldsymbol{a}' = \begin{bmatrix} 0.3536 \\ -0.4154 \end{bmatrix}, \ \boldsymbol{b}' = \begin{bmatrix} -0.35356 \\ -0.4154 \end{bmatrix}, \ \boldsymbol{c}' = \begin{bmatrix} 0 \\ 0 \end{bmatrix}$$

from which matrices $\boldsymbol{P}$ and $\boldsymbol{P}'$ are readily constructed. In the sequel, we will also need $\boldsymbol{P}^{-1}$. These three matrices are displayed below[3]:

$$\boldsymbol{P} = \begin{bmatrix} 1 & 0 & 0.7937 \\ 0 & 1 & 0.7937 \\ 1 & 1 & 1 \end{bmatrix}, \ \boldsymbol{P}' = \begin{bmatrix} 0.3536 & -0.3536 & 0 \\ -0.4154 & -0.4154 & 0 \\ 1 & 1 & 1 \end{bmatrix}, \ \boldsymbol{P}^{-1} = \begin{bmatrix} -0.3512 & -1.351 & 1.351 \\ -1.351 & -0.3512 & 1.351 \\ 1.702 & 1.702 & -1.702 \end{bmatrix}$$

The homogeneous affine transformation matrix sought is, thus,

$$\boldsymbol{T} = \boldsymbol{P}'\boldsymbol{P}^{-1} = \begin{bmatrix} 0.3536 & -0.3536 & 0 \\ 0.7071 & 0.7071 & -1.122 \\ 0 & 0 & 1 \end{bmatrix}$$

its inverse being also a homogeneous-transformation matrix, namely,

[3]The inverse of $\boldsymbol{P}$ was obtained with computer algebra.

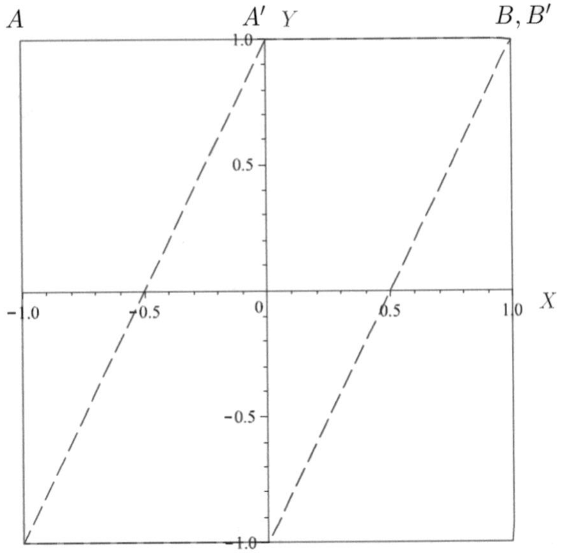

Fig. 4.13 The distortion of a square into a parallelogram by means of an affine transformation known as *shear*

$$T^{-1} = \begin{bmatrix} 1.414 & 0.7071 & 0.7937 \\ -1.414 & 0.7071 & 0.7937 \\ 0 & 0 & 1 \end{bmatrix}$$

thereby completing the example.

Example 4.2.4 (*Shear*) *Shear* is sometimes referred to as one more case of simple affine transformation, not obtained from a combination of the four studied in Sect. 4.1: scaling, translation, rotation, and reflection. In fact, shear can be obtained as a concatenation of those four simple transformations, as shown here. To this end, find the affine transformation mapping the square of Fig. 4.13 into the parallelogram shown dashed in the same figure.

Solution Let A and B denote the upper left and right corners of the square, with $\boldsymbol{a}$ and $\boldsymbol{b}$ denoting their position vectors. Their images, A' and B', are the corresponding corners of the parallelogram, i.e.,

$$\boldsymbol{a} = \begin{bmatrix} -1 \\ 1 \end{bmatrix}, \quad \boldsymbol{b} = \begin{bmatrix} 1 \\ 1 \end{bmatrix}, \quad \boldsymbol{a}' = \begin{bmatrix} 0 \\ 1 \end{bmatrix}, \quad \boldsymbol{b}' = \begin{bmatrix} 1 \\ 1 \end{bmatrix}$$

matrices $\boldsymbol{P}$ and $\boldsymbol{P}'$, as well as the inverse of the former, as given below:

$$\boldsymbol{P} = \begin{bmatrix} -1 & 1 \\ 1 & 1 \end{bmatrix}, \quad \boldsymbol{P}' = \begin{bmatrix} 0 & 1 \\ 1 & 1 \end{bmatrix}, \quad \boldsymbol{P}^{-1} = \frac{1}{2} \begin{bmatrix} -1 & 1 \\ 1 & 1 \end{bmatrix}$$

The transformation matrix $\boldsymbol{M}$ and the homogeneous-transformation matrix $\boldsymbol{T}$ are hence readily obtained:

$$\boldsymbol{M} = \boldsymbol{P}'\boldsymbol{P}^{-1} = \begin{bmatrix} 1/2 & 1/2 \\ 0 & 1 \end{bmatrix}, \quad \boldsymbol{T} = \begin{bmatrix} 1/2 & 1/2 & 0 \\ 0 & 1 & 0 \\ 0 & 0 & 1 \end{bmatrix}$$

The *upper triangular* form of $\boldsymbol{M}$ is to be highlighted. This form is characteristic of shear transformations. Furthermore, matrix $\boldsymbol{M}$ can be factored[4] into the form

$$\boldsymbol{M} = \boldsymbol{QSE}$$

where $\boldsymbol{Q}$ and $\boldsymbol{E}$ are, in general, orthogonal matrices, proper or improper, while $\boldsymbol{S}$ is a simple scaling matrix of the form displayed in Eq. (4.4). In our case, both $\boldsymbol{Q}$ and $\boldsymbol{E}$ turn out to be proper orthogonal, and hence represent rotations. All three factors are displayed below:

$$\boldsymbol{Q} = \begin{bmatrix} 0.5257 & -0.8507 \\ 0.8507 & 0.5257 \end{bmatrix}, \quad \boldsymbol{S} = \begin{bmatrix} 1.144 & 0 \\ 0 & 0.4370 \end{bmatrix}, \quad \boldsymbol{E} = \begin{bmatrix} 0.2298 & -0.9732 \\ 0.9732 & 0.2298 \end{bmatrix}$$

Moreover, $\boldsymbol{Q}$ represents a rotation about the origin through an angle of 58.28° ccw, while $\boldsymbol{E}$ a rotation about the origin as well, of 76.72° ccw. The distortion of the square can thus be regarded as a simple scaling of values $S_{x'} = 1.144$, $S_{y'} = 0.4370$ about axes X'–Y' rotated ccw through an angle of 76.72°, followed by a rotation of the distorted figure through an angle of 58.28° ccw.

If the same transformation $\boldsymbol{T}$ is applied to the unit circle centered at the origin, the circle image is an ellipse of semiaxes $S_{x'}$, $S_{y'}$, its major axis at an angle of 76.72° ccw with the X-axis, as shown in Fig. 4.14.

4.2.2 *Affine Transformations of 2D Objects Under Implicit Representations*

The implementation of affine transformations given above applies to *parametric representations* of the objects at hand. Sometimes, these representations exhibit certain limitations. For example, while Lamé curves of even degree are closed and those of odd degree are unbounded, their parametric representations apply only to one restricted portion of the curve, namely, that for which the parameter, an angle, attains values between 0 and 90°. This limits the parametric representation of Lamé

[4]The factoring is based on the *polar decomposition* (PD) of $\boldsymbol{M}$ and the *eigenvalue decomposition* of the symmetric, positive-definite factor of the PD. These two decompositions lying outside of the scope of the book, they are not discussed here. This factoring is related to, but simpler than, the most popular *singular-value decomposition* of matrix $\boldsymbol{M}$: Strang (1986).

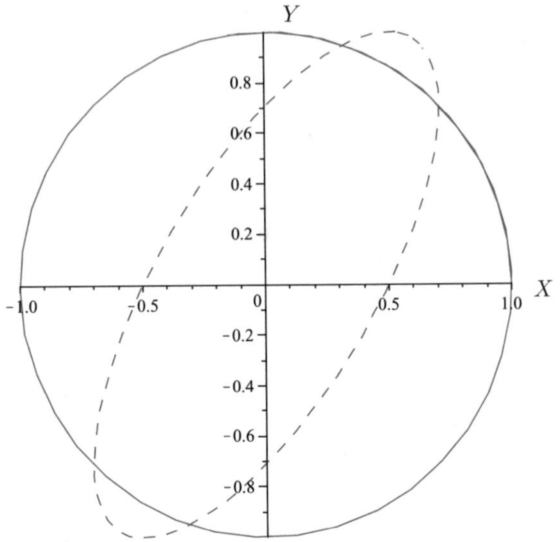

Fig. 4.14 A shear of the unit circle centered at the origin that carries it into an ellipse with its major axis making an angle of about 58.28° with the X-axis ccw

curves to the first quadrant. The cubic Lamé curve and its distorted and displaced image of Fig. 4.11 were obtained using the implicit representation of Eq. (2.42). For this reason, it is convenient to study the application of affine transformations to planar objects described by implicit representations.

Let us assume that the ellipse $\mathcal{E}$ of semiaxes 2 and 1, centered at the origin, is to be scaled by factors $S_x = 2$, $S_y = 1$. The implicit equation of the ellipse is given by

$$\mathcal{E}: \quad \frac{x^2}{4} + y^2 - 1 = 0$$

Application of the given scaling to the position vector $\boldsymbol{p} = [\,x,\ y\,]^T$ of a point on the ellipse yields

$$\boldsymbol{q} = \begin{bmatrix} 2 & 0 \\ 0 & 1 \end{bmatrix} \begin{bmatrix} x \\ y \end{bmatrix} = \begin{bmatrix} 2x \\ y \end{bmatrix}$$

Now, let us replace the coordinates of the new point Q, of position vector $\boldsymbol{q}$, into the equation $\mathcal{E}$ of the ellipse:

$$\frac{4x^2}{4} + y^2 - 1 = 0 \quad \text{or} \quad x^2 + y^2 = 1$$

which is the equation of the unit circle centered at the origin. Not what we intended to obtain! If we apply, instead, the inverse scaling, i.e., one with $S_x = 1/2$, $S_y = 1$, then

$$\boldsymbol{q} = \begin{bmatrix} 1/2 & 0 \\ 0 & 1 \end{bmatrix} \begin{bmatrix} x \\ y \end{bmatrix} = \begin{bmatrix} x/2 \\ y \end{bmatrix}$$

Now, the coordinates of the new point Q are substituted into the original equation $\mathcal{E}$ of the ellipse, to obtain

$$\frac{x^2}{16} + y^2 - 1 = 0$$

which is, indeed, the equation of an ellipse of semiaxes 4 and 1, as desired.

Furthermore, let us now attempt a rotation of the same original ellipse $\mathcal{E}$ through an angle of 30° ccw. To this end, the proper orthogonal matrix $\boldsymbol{R}$ is introduced that represents a rotation about the origin through an angle of 30° ccw:

$$\boldsymbol{R} = \begin{bmatrix} \sqrt{3}/2 & -1/2 \\ 1/2 & \sqrt{3}/2 \end{bmatrix}$$

which, when applied to vector $\boldsymbol{p}$ given above, yields a rotated vector $\boldsymbol{r}$, namely,

$$\boldsymbol{r} = \boldsymbol{R}\boldsymbol{p} = \begin{bmatrix} \sqrt{3}/2 & -1/2 \\ 1/2 & \sqrt{3}/2 \end{bmatrix} \begin{bmatrix} x \\ y \end{bmatrix} = \begin{bmatrix} \sqrt{3}x/2 - y/2 \\ x/2 + \sqrt{3}y/2 \end{bmatrix} \equiv \begin{bmatrix} r_1 \\ r_2 \end{bmatrix}$$

Upon substituting the components of this vector into the equation of the ellipse, with r_1 replacing x and r_2 y, we obtain the equation of the rotated ellipse:

$$\frac{7}{16}x^2 + \frac{3}{8}\sqrt{3}xy + \frac{13}{16}y^2 = 1$$

which, alas, is not what we wanted. The reader is invited to plot the foregoing equation to find that the ellipse it represents is the original one rotated through an angle of 30° cw, rather than ccw. To remedy this effect, the rule is quite simple: if a 2D curve is given in *implicit form*, and it is desired to apply an affine transformation $\boldsymbol{T}$ to this curve, apply to a generic point of the curve the transformation $\boldsymbol{T}^{-1}$ and substitute the new coordinates in the implicit equation of the curve of interest.

What the above experience shows is the effect of rotating an object while leaving the coordinate axes fixed. This rotation is equivalent to the effect of rotating the axes in the opposite direction by the same amount, while leaving the object fixed. This observation is made apparent in Fig. 4.4: Rotating the McGill crest through +45°—i.e., ccw—while leaving the X-, Y-axes fixed is equivalent to rotating the axes through −45°—i.e., cw—while leaving the McGill crest fixed.

As a matter of fact, the foregoing equivalence applies to *any* affine transformation. That is, when plotting the image of an object represented by an implicit equation and undergoing a given affine transformation, the position vector of any of its points has to be multiplied by the *inverse* of the given transformation.

Example 4.2.5 Given the cubic Lamé curve of Fig. 2.17b, produce a rendering of the same curve, with its "bump" B located at the origin of the X–Y-plane and its tangent at the same point coinciding with the X-axis.

Solution Apparently, a general displacement of the curve is needed, involving both a rotation $\boldsymbol{R}$ and a translation $\boldsymbol{t}$. The rotation is through an angle of 45° ccw about the origin, the translation toward the origin by an amount d from B. The total displacement can be represented in one single 3×3 homogeneous-transformation matrix $\boldsymbol{T}_d$. The value of d was found in Example 4.2.3 to be $d = \sqrt{2}a$, with $a = (1/2)^{1/3}$. Apparently, what we need is a displacement $\boldsymbol{T}_d$ involving (a) a rotation of +45° (ccw) and (b) with the curve already rotated, a translation $\boldsymbol{t} = [0, \ -d]^T$:

$$\boldsymbol{T}_d = \begin{bmatrix} \boldsymbol{R} & \boldsymbol{t} \\ \boldsymbol{0}^T & 1 \end{bmatrix} = \begin{bmatrix} \sqrt{2}/2 & -\sqrt{2}/2 & 0 \\ \sqrt{2}/2 & \sqrt{2}/2 & -2^{1/6} \\ 0 & 0 & 1 \end{bmatrix}$$

In order to implement the transformation of the implicit equation,

$$\mathcal{L}_3: \quad x^3 + y^3 - 1 = 0$$

we need the inverse of $\boldsymbol{T}_d$, which is readily computed using the general expression of Eq. (1.121):

$$\boldsymbol{T}_d^{-1} = \begin{bmatrix} \sqrt{2}/2 & \sqrt{2}/2 & 2^{-1/3} \\ -\sqrt{2}/2 & \sqrt{2}/2 & 2^{-1/3} \\ 0 & 0 & 1 \end{bmatrix}$$

Let $\boldsymbol{p} = [x, \ y, \ 1]^T$ be the vector of homogeneous coordinates of a point P on $\mathcal{L}_3$, its image under $\boldsymbol{T}_d$ being $\boldsymbol{q}$, i.e.,

$$\boldsymbol{q} = \boldsymbol{T}_d^{-1}\boldsymbol{p} = \begin{bmatrix} \sqrt{2}/2 & \sqrt{2}/2 & 2^{-1/3} \\ -\sqrt{2}/2 & \sqrt{2}/2 & 2^{-1/3} \\ 0 & 0 & 1 \end{bmatrix} \begin{bmatrix} x \\ y \\ 1 \end{bmatrix} = \begin{bmatrix} (\sqrt{2}x)/2 + (\sqrt{2}y)/2 + 2^{-1/3} \\ -(\sqrt{2}x)/2 + (\sqrt{2}y)/2 + 2^{-1/3} \\ 1 \end{bmatrix}$$

Both the original cubic Lamé curve and its displaced image are shown in Fig. 4.15.

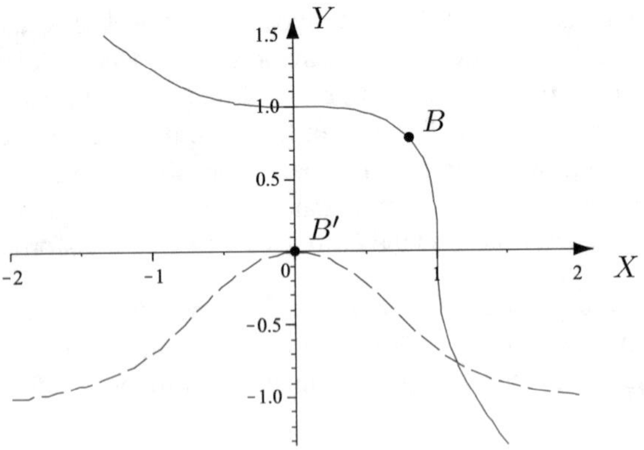

Fig. 4.15 A cubic Lamé curve and its image after a translation and a rotation

4.3 3D Transformations

Transformations in three-dimensional space are executed by the same methods used in two-dimensional space, with the addition of the z-coordinate. In homogeneous coordinates, these transformations are represented by a 4×4 homogeneous transformation matrix $\boldsymbol{T}$, mapping the four-dimensional array $\boldsymbol{p}$ of homogeneous coordinates of P into its counterpart $\boldsymbol{p}'$ of P' in the form

$$\boldsymbol{p}' = \boldsymbol{T}\boldsymbol{p} \tag{4.30a}$$

where

$$\boldsymbol{T} = \begin{bmatrix} A & B & C & D \\ E & F & G & H \\ I & J & K & L \\ 0 & 0 & 0 & 1 \end{bmatrix}, \quad \boldsymbol{p} = \begin{bmatrix} x \\ y \\ z \\ 1 \end{bmatrix}, \quad \boldsymbol{p}' = \begin{bmatrix} x' \\ y' \\ z' \\ 1 \end{bmatrix} \tag{4.30b}$$

Moreover, matrix $\boldsymbol{T}$ is partitioned in the usual form:

$$\boldsymbol{T} = \begin{bmatrix} \boldsymbol{M} & \boldsymbol{t} \\ \boldsymbol{0}^T & 1 \end{bmatrix} \tag{4.31}$$

The 3×3 matrix $\boldsymbol{M}$ in the upper left corner allows for scaling, reflection, and rotation, while the three-dimensional vector $\boldsymbol{t}$ accounts for the translation. Lastly, $\boldsymbol{0}$ is the three-dimensional zero vector.

4.3.1 Scaling

The scaling transformation is obtained by placing values along the main diagonal of the general 4×4 transformation matrix. The coordinates of an arbitrary point $P(x, y, z, 1)$ are scaled into those of $P'(x', y', z', 1)$ by means of the transformation

$$\begin{bmatrix} x' \\ y' \\ z' \\ 1 \end{bmatrix} = \begin{bmatrix} A & 0 & 0 & 0 \\ 0 & F & 0 & 0 \\ 0 & 0 & K & 0 \\ 0 & 0 & 0 & 1 \end{bmatrix} \begin{bmatrix} x \\ y \\ z \\ 1 \end{bmatrix} \tag{4.32a}$$

In this case, matrix $\boldsymbol{M}$ is diagonal, while $\boldsymbol{t} = \boldsymbol{0}$, for there is no translation involved. Matrix $\boldsymbol{M}$ is, thus,

$$\boldsymbol{M} = \mathrm{diag}(A, F, K) \tag{4.32b}$$

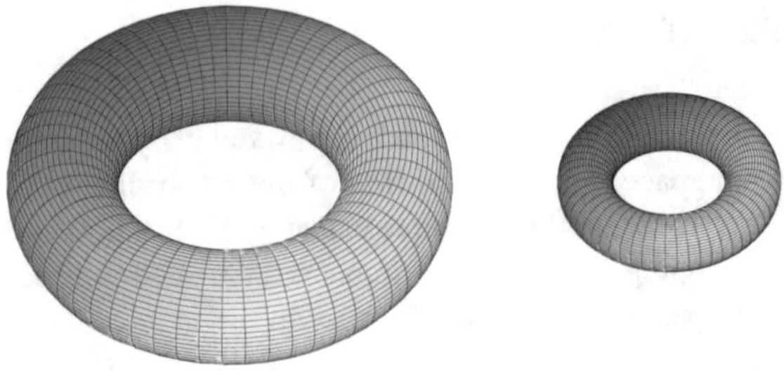

Fig. 4.16 Uniform scaling

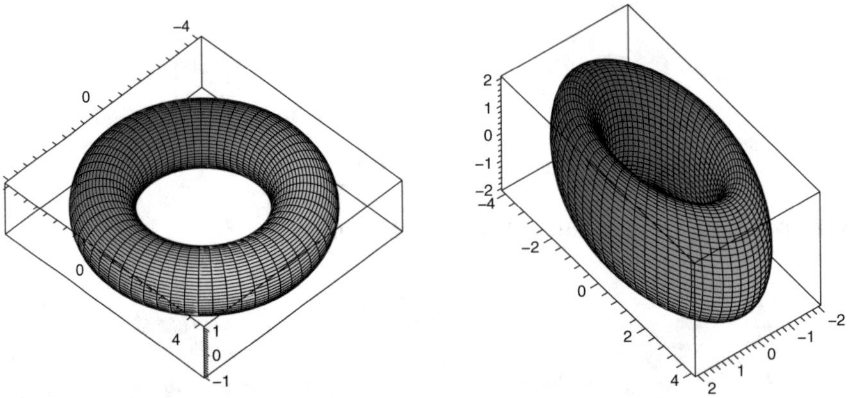

Fig. 4.17 Non-uniform scaling in 3D

This is an extension of two-dimensional scaling, described in Sect. 4.1.1. If the scaling factors A, F, K are not equal, the image of the object is distorted. Otherwise, a change in size occurs, but the original proportions are maintained.

In Fig. 4.16, a torus is uniformly scaled to form the smaller torus. How to construct a torus is explained in Sect. 4.5.1.

In Fig. 4.17, the torus on the left is scaled by $(0.5,\ 1,\ 2)$ in the $(x,\ y,\ z)$ directions, respectively, thus producing the surface of the right, which is no longer a surface of revolution. Notice that the central circle[5] of the torus becomes an ellipse of semiaxes of length ratio 0.5, while the cross section obtained by cutting the object by a horizontal plane a distance h from the base determines two ellipses, the smaller inside the larger one, of semiaxes length ratios depending on h.

[5]This is the circle traced by the center of the circle playing the role of the generatrix of the torus, which is produced by a rotation of the circle about the axis of the torus.

Fig. 4.18 Translations in 3D

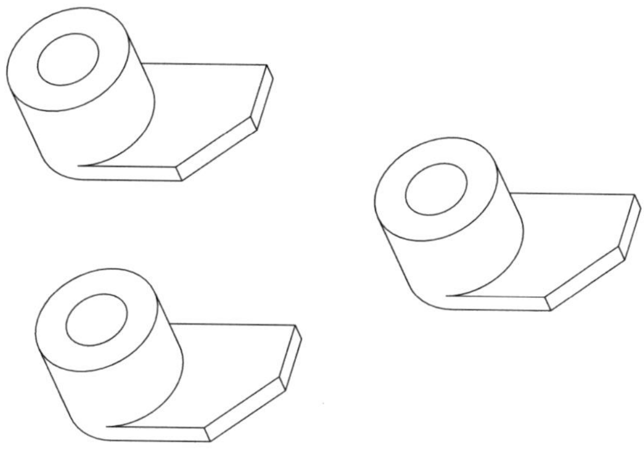

4.3.2 Translation

A translation is a special case of *rigid-body displacement*, under which all the points of the body undergo the same displacement. The transformation translating a point $P(x, y, z, 1)$ to a new point $P'(x', y', z', 1)$ through a prescribed displacement $\boldsymbol{t}$ is given by

$$\begin{bmatrix} x' \\ y' \\ z' \\ 1 \end{bmatrix} = \begin{bmatrix} 1 & 0 & 0 & D \\ 0 & 1 & 0 & H \\ 0 & 0 & 1 & L \\ 0 & 0 & 0 & 1 \end{bmatrix} \begin{bmatrix} x \\ y \\ z \\ 1 \end{bmatrix} \tag{4.33a}$$

Notice that, in this case, all the points of the body undergo the same displacement, but the object is neither rotated nor distorted. We thus have

$$\boldsymbol{M} = \boldsymbol{1}, \quad \boldsymbol{t} = \begin{bmatrix} D & H & L \end{bmatrix}^T \tag{4.33b}$$

The values of D, H, L represent the relative translation of the point in the x-, y-, z-directions, respectively.

In Fig. 4.18, we can see examples of translations.

4.3.3 Rotation

A rotation in 3D is another special case of a rigid-body displacement. Under such a rotation, the distance between every pair of object points is preserved, with the distance of every point of the object to a line $\mathcal{L}$, either within the body or outside of it, remaining constant. The object is said to rotate about $\mathcal{L}$, the *axis of rotation*.

Rotations in three dimensions are more complex to describe and to represent than their two-dimensional counterparts, because an *axis of rotation*, $\mathcal{L}$, rather than a center of rotation, must be specified. Rotations about an axis passing through the origin are characterized by a *proper orthogonal* matrix $\boldsymbol{M}$ and a zero translation, $\boldsymbol{t} = \boldsymbol{0}$. In this case, $\boldsymbol{M}$ has the properties below:

$$\boldsymbol{M}^T\boldsymbol{M} = \boldsymbol{M}\boldsymbol{M}^T = \boldsymbol{1}, \quad \det(\boldsymbol{M}) = +1 \tag{4.34}$$

In particular, the matrix representing a rotation about the Z-axis through an angle θ is

$$\boldsymbol{M}_Z = \begin{bmatrix} \cos\theta & -\sin\theta & 0 \\ \sin\theta & \cos\theta & 0 \\ 0 & 0 & 1 \end{bmatrix} \tag{4.35}$$

which produces the mapping:

$$\begin{aligned} x' &= x\cos\theta - y\sin\theta \\ y' &= x\sin\theta + y\cos\theta \\ z' &= z \end{aligned} \tag{4.36}$$

In a similar manner, a rotation of θ about the Y-axis can be obtained by means of

$$\begin{aligned} x' &= x\cos\theta + z\sin\theta \\ y' &= y \\ z' &= -x\sin\theta + z\cos\theta \end{aligned} \tag{4.37}$$

and is correspondingly represented by

$$\boldsymbol{M}_Y = \begin{bmatrix} \cos\theta & 0 & \sin\theta \\ 0 & 1 & 0 \\ -\sin\theta & 0 & \cos\theta \end{bmatrix} \tag{4.38}$$

A rotation about the X-axis is

$$\begin{aligned} x' &= x \\ y' &= y\cos\theta - z\sin\theta \\ z' &= y\sin\theta + z\cos\theta \end{aligned} \tag{4.39}$$

which is represented by

$$\boldsymbol{M}_X = \begin{bmatrix} 1 & 0 & 0 \\ 0 & \cos\theta & -\sin\theta \\ 0 & \sin\theta & \cos\theta \end{bmatrix} \tag{4.40}$$

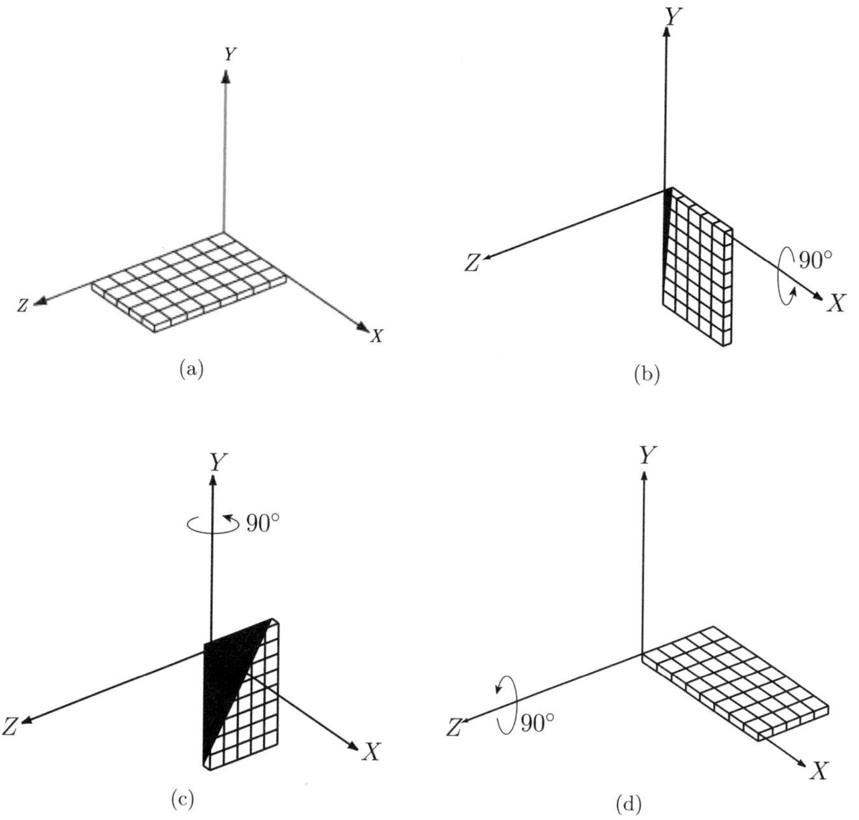

Fig. 4.19 A solar panel: **a** in its original configuration; **b** followed by a rotation through 90° about the X-axis; **c** then by a second rotation through 90° about the Y-axis; and **d** finally, a third rotation through 90° about the Z-axis

Sometimes, rotations about arbitrary axes are specified as a sequence of rotations about the coordinate axes, as illustrated with the solar panel of Fig. 4.19a, used in telecommunication satellites to provide energy to their different instruments. In this case, we have

$$\boldsymbol{M}_X = \begin{bmatrix} 1 & 0 & 0 \\ 0 & 0 & -1 \\ 0 & 1 & 0 \end{bmatrix}, \quad \boldsymbol{M}_Y = \begin{bmatrix} 0 & 0 & 1 \\ 0 & 1 & 0 \\ -1 & 0 & 0 \end{bmatrix}, \quad \boldsymbol{M}_Z = \begin{bmatrix} 0 & -1 & 0 \\ 1 & 0 & 0 \\ 0 & 0 & 1 \end{bmatrix}$$

Example 4.3.1 With reference to Fig. 4.19, (a) find the homogeneous-transformation matrix that carries the solar panel from attitude (a) to attitude (c) and takes its point P that coincides with the origin, in attitude (a), to a new location, labeled $Q(1, 2, 3)$; then (b) find the homogeneous-transformation matrix that will carry the same panel back to its original configuration.

Solution

(a) Let $\boldsymbol{M}_{ac}$ be the matrix representing the rotation from attitude (a) to attitude (c). Thus,

$$\boldsymbol{M}_{ac} = \boldsymbol{M}_Y \boldsymbol{M}_X = \begin{bmatrix} 0 & 1 & 0 \\ 0 & 0 & -1 \\ -1 & 0 & 0 \end{bmatrix}$$

the corresponding homogeneous transformation $\boldsymbol{T}_r$ being readily obtained as

$$\boldsymbol{T}_r = \begin{bmatrix} 0 & 1 & 0 & 0 \\ 0 & 0 & -1 & 0 \\ -1 & 0 & 0 & 0 \\ 0 & 0 & 0 & 1 \end{bmatrix}$$

Now we need a homogeneous transformation $\boldsymbol{T}_t$ to translate the panel without rotating it. Given the translation vector $\boldsymbol{t}$ involved, obtaining $\boldsymbol{T}_t$ is straightforward:

$$\boldsymbol{T}_t = \begin{bmatrix} 1 & 0 & 0 & 1 \\ 0 & 1 & 0 & 2 \\ 0 & 0 & 1 & 3 \\ 0 & 0 & 0 & 1 \end{bmatrix}$$

Hence, the total transformation matrix $\boldsymbol{T}$ is given by

$$\boldsymbol{T} = \boldsymbol{T}_t \boldsymbol{T}_r = \begin{bmatrix} \boldsymbol{1} & \boldsymbol{t} \\ \boldsymbol{0}^T & 1 \end{bmatrix} \begin{bmatrix} \boldsymbol{M}_{ac} & \boldsymbol{0} \\ \boldsymbol{0}^T & 1 \end{bmatrix} = \begin{bmatrix} \boldsymbol{M}_{ac} & \boldsymbol{t} \\ \boldsymbol{0}^T & 1 \end{bmatrix}$$

i.e.,

$$\boldsymbol{T} = \begin{bmatrix} 0 & 1 & 0 & 1 \\ 0 & 0 & -1 & 2 \\ -1 & 0 & 0 & 3 \\ 0 & 0 & 0 & 1 \end{bmatrix}$$

(b) To compute $\boldsymbol{T}^{-1}$, we recall the general expression (1.121) for the inverse of a homogeneous transformation matrix:

$$\boldsymbol{T}^{-1} = \begin{bmatrix} \boldsymbol{M}_{ac}^{-1} & -\boldsymbol{M}_{ac}^{-1}\boldsymbol{t} \\ \boldsymbol{0}^T & 1 \end{bmatrix} = \begin{bmatrix} \boldsymbol{M}_{ac}^T & -\boldsymbol{M}_{ac}^T\boldsymbol{t} \\ \boldsymbol{0}^T & 1 \end{bmatrix}$$

where the orthogonality of $\boldsymbol{M}_{ac}$ has been recalled, to simplify the foregoing expression, the final result thus being

$$T^{-1} = \begin{bmatrix} 0 & 0 & -1 & 3 \\ 1 & 0 & 0 & -1 \\ 0 & -1 & 0 & 2 \\ 0 & 0 & 0 & 1 \end{bmatrix}$$

thereby completing the calculations required.

Example 4.3.2 Matrix $\boldsymbol{M}$ shown below is claimed to represent a rotation of an object $\mathcal{B}$ about a given axis passing through the origin.

$$\boldsymbol{M} = \begin{bmatrix} 0\,1\,0 \\ 0\,0\,1 \\ 1\,0\,0 \end{bmatrix}$$

(a) Prove that the matrix indeed represents a rotation; then
(b) Find its axis and its angle of rotation

Solution

(a) To represent a rotation, $\boldsymbol{M}$ must be proper orthogonal. We thus compute

$$\boldsymbol{M}\boldsymbol{M}^T = \begin{bmatrix} 0\,1\,0 \\ 0\,0\,1 \\ 1\,0\,0 \end{bmatrix} \begin{bmatrix} 0\,0\,1 \\ 1\,0\,0 \\ 0\,1\,0 \end{bmatrix} = \begin{bmatrix} 1\,0\,0 \\ 0\,1\,0 \\ 0\,0\,1 \end{bmatrix}$$

or, equivalently

$$\boldsymbol{M}^T\boldsymbol{M} = \begin{bmatrix} 0\,0\,1 \\ 1\,0\,0 \\ 0\,1\,0 \end{bmatrix} \begin{bmatrix} 0\,1\,0 \\ 0\,0\,1 \\ 1\,0\,0 \end{bmatrix} = \begin{bmatrix} 1\,0\,0 \\ 0\,1\,0 \\ 0\,0\,1 \end{bmatrix}$$

and hence $\boldsymbol{M}$ is indeed orthogonal. To be proper orthogonal, its determinant must be $+1$. We thus compute its determinant by expansion of cofactors of its first column:

$$\det(\boldsymbol{M}) = 0 + 0 + \det \begin{bmatrix} 1\,0 \\ 0\,1 \end{bmatrix} = +1$$

thereby showing that $\boldsymbol{M}$ indeed represents a rotation.

(b) As will become apparent presently, to find the axis of rotation of $\boldsymbol{M}$, we need the initial and final positions of two points of $\mathcal{B}$, none of which is the origin, for the origin does not move—it is a point of the axis of rotation! Let us thus choose point A a distance d_A from the origin in the positive direction of X, and a second point B a distance d_B in the positive direction of Y, as shown in Fig. 4.20.

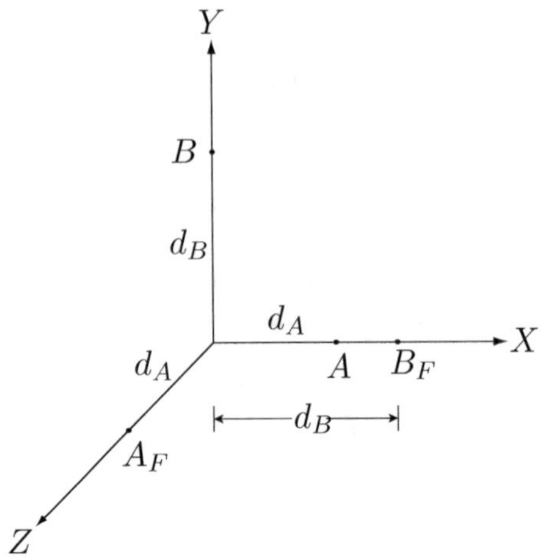

Fig. 4.20 Two points, A and B, of an object $\mathcal{B}$ rotating about an axis that passes through the origin, in the original and the final attitudes of $\mathcal{B}$, with the final point positions carrying the subscript F

Let, moreover, $\boldsymbol{a}$ and $\boldsymbol{b}$ be the position vectors of A and B, respectively, i.e.,

$$\boldsymbol{a} = \begin{bmatrix} d_A \\ 0 \\ 0 \end{bmatrix}, \quad \boldsymbol{b} = \begin{bmatrix} 0 \\ d_B \\ 0 \end{bmatrix}$$

In the final attitude of $\mathcal{B}$, A and B take the positions A_F and B_F, respectively, of position vectors $\boldsymbol{a}_F$ and $\boldsymbol{b}_F$. That is,

$$\boldsymbol{a}_F = \boldsymbol{M}\boldsymbol{a} = \begin{bmatrix} 0 & 1 & 0 \\ 0 & 0 & 1 \\ 1 & 0 & 0 \end{bmatrix} \begin{bmatrix} d_A \\ 0 \\ 0 \end{bmatrix} = \begin{bmatrix} 0 \\ 0 \\ d_A \end{bmatrix}$$

$$\boldsymbol{b}_F = \boldsymbol{M}\boldsymbol{b} = \begin{bmatrix} 0 & 1 & 0 \\ 0 & 0 & 1 \\ 1 & 0 & 0 \end{bmatrix} \begin{bmatrix} 0 \\ d_B \\ 0 \end{bmatrix} = \begin{bmatrix} d_B \\ 0 \\ 0 \end{bmatrix}$$

The axis of rotation is the set of points of $\mathcal{B}$ that do not change their position in the final attitude of the body. These points are known[6] to lie in a line $\mathcal{L}$ passing through the origin, in the direction of the unit vector $\boldsymbol{e}$, as illustrated in Fig. 4.21. Since a rotation entails no distortion, all points of $\mathcal{L}$ are equidistant from P and P_F. In particular, Q is the intersection of the perpendiculars to $\mathcal{L}$ from P and P_F. Angle ϕ, measured as indicated in Fig. 4.21, is the angle of rotation. Notice that if vector $\boldsymbol{e}$ is defined as pointing in the opposite direction, then ϕ reverses

[6]Leonhard Euler proved this in 1776.

Fig. 4.21 An object $\mathcal{B}$ in its original and final attitudes; illustration of axis $\mathcal{L}$ and angle of rotation ϕ

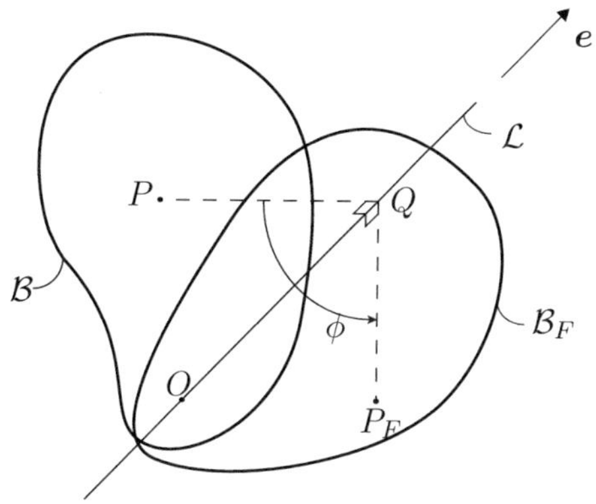

its sign.

Apparently, $\mathcal{L}$ lies in the *bisector plane*[7] of segment $\overline{PP_F}$. Hence, to find $\mathcal{L}$, all we need is two of its points in their original and final positions. The intersection of the two bisector planes of the segments defined by these points then gives $\mathcal{L}$. We thus start by finding the bisector planes Π_A and Π_B of segments $\overline{AA_F}$ and $\overline{BB_F}$, respectively. To this end, we recall Eq. (3.2), thus obtaining

$$\Pi_A : \quad (\boldsymbol{a}_F - \boldsymbol{a})^T \boldsymbol{p} + \frac{1}{2}(||\boldsymbol{a}||^2 - ||\boldsymbol{a}_F||^2) = 0$$

$$\Pi_B : \quad (\boldsymbol{b}_F - \boldsymbol{b})^T \boldsymbol{p} + \frac{1}{2}(||\boldsymbol{b}||^2 - ||\boldsymbol{b}_F||^2) = 0$$

In our case,

$$||\boldsymbol{a}||^2 = ||\boldsymbol{a}_F||^2 = d_A^2, \quad ||\boldsymbol{b}||^2 = ||\boldsymbol{b}_F||^2 = d_B^2$$

$$\boldsymbol{a}_F - \boldsymbol{a} = \begin{bmatrix} -d_A \\ 0 \\ d_A \end{bmatrix}, \quad \boldsymbol{b}_F - \boldsymbol{b} = \begin{bmatrix} d_B \\ -d_B \\ 0 \end{bmatrix}$$

Hence, the equations of the two planes are

$$\Pi_A : \quad -d_A x + d_A z = 0$$

$$\Pi_B : \quad d_B x - d_B y = 0$$

which gives a homogeneous system of two linear equations in three unknowns. These two equations do define a line $\mathcal{L}$ passing through the origin—the coordinates of the origin, (0,0,0), verify the two equations—which is nothing but the

[7]The bisector plane of two given points is the locus of points equidistant from the two points.

axis of rotation sought.
To compute the direction cosines of $\mathcal{L}$, (λ, μ, ν), all we need is substitute

$$\lambda \leftarrow x, \quad \mu \leftarrow y, \quad \nu \leftarrow z$$

in the two above equations, and impose the condition that the sum of the squares of the three direction cosines be unity, i.e.,

$$\begin{aligned} -\lambda + \nu &= 0 \\ \lambda - \mu &= 0 \\ \lambda^2 + \mu^2 + \nu^2 &= 1 \end{aligned}$$

From the first and the second equations, we obtain

$$\nu = \lambda, \quad \mu = \lambda$$

the third equation thus yielding

$$3\lambda^2 = 1 \quad \Rightarrow \lambda = \pm\frac{\sqrt{3}}{3}$$

Picking up the positive sign, for example, we obtain

$$\lambda = \mu = \nu = \frac{\sqrt{3}}{3} \quad \Rightarrow \boldsymbol{e} = \frac{\sqrt{3}}{3}\begin{bmatrix} 1 \\ 1 \\ 1 \end{bmatrix}$$

the last expression following from the realization that the direction cosines of $\mathcal{L}$ are nothing but the components of the unit vector $\boldsymbol{e}$, giving the direction of $\mathcal{L}$, thereby finding the axis of rotation. Further, to find the angle of rotation, we need a point Q on $\mathcal{L}$ that lies on the normal to $\mathcal{L}$ from, say A—B might as well be taken, without affecting the final result—so that $\overline{QA}$ is normal to $\mathcal{L}$. If we let (ξ, η, ζ) be the coordinates of Q, then the foregoing normality condition can be expressed as

$$\boldsymbol{e}^T(\boldsymbol{q} - \boldsymbol{a}) = 0 \quad \Rightarrow \frac{\sqrt{3}}{3}[1\ 1\ 1]\begin{bmatrix} \xi - d_A \\ \eta \\ \zeta \end{bmatrix} = 0$$

where $\boldsymbol{q}$ is the position vector of Q. Hence,

$$\xi - d_A + \eta + \zeta = 0$$

thereby obtaining one equation for the three coordinates of Q. Two more equations are available if we realize that Q is a point of $\mathcal{L}$, and hence its coordinates verify the equations of Π_A and Π_B:

$$-\xi + \zeta = 0, \quad \xi - \eta = 0$$

The two above equations yield $\eta = \xi$, $\zeta = \xi$. When these expressions are substituted in the foregoing normality condition, we obtain

$$3\xi = d_A \quad \Rightarrow \xi = \frac{1}{3} d_A$$

Hence,

$$\boldsymbol{q} = \begin{bmatrix} d_A/3 \; d_A/3 \; d_A/3 \end{bmatrix}^T$$

Now, ϕ can be obtained from the relation

$$(\boldsymbol{a} - \boldsymbol{q})^T (\boldsymbol{a}_F - \boldsymbol{q}) = ||\boldsymbol{a} - \boldsymbol{q}||^2 \cos\phi$$

where

$$\boldsymbol{a} - \boldsymbol{q} = \begin{bmatrix} 2d_A/3 \\ -d_A/3 \\ -d_A/3 \end{bmatrix}, \quad \boldsymbol{a}_F - \boldsymbol{q} = \begin{bmatrix} -d_A/3 \\ -d_A/3 \\ 2d_A/3 \end{bmatrix}, \quad ||\boldsymbol{a} - \boldsymbol{q}||^2 = \frac{2}{3} d_A^2$$

Therefore,

$$\frac{d_A^2}{9}(-2 + 1 - 2) = \frac{2}{3} d_A^2 \cos\phi, \quad \Rightarrow \cos\phi = -\frac{1}{2}$$

from which $\phi = 120°$ or $240°$. To destroy the ambiguity, we take into account the direction given by $\boldsymbol{e}$. Hence,

$$(\boldsymbol{a} - \boldsymbol{q}) \times (\boldsymbol{a}_F - \boldsymbol{q}) = (\sin\phi)||\boldsymbol{a} - \boldsymbol{q}||^2 \boldsymbol{e}$$

If we dot multiply both sides of the above equation by $\boldsymbol{e}$, we obtain an equation for $\sin\phi$:

$$||\boldsymbol{a} - \boldsymbol{q}||^2 \sin\phi = (\boldsymbol{a} - \boldsymbol{q}) \times (\boldsymbol{a}_F - \boldsymbol{q}) \cdot \boldsymbol{e} = \det([\boldsymbol{a} - \boldsymbol{q}, \boldsymbol{a}_F - \boldsymbol{q}, \boldsymbol{e}])$$

Therefore,

$$\frac{2}{3} d_A^2 \sin\phi = \det \begin{bmatrix} \frac{2d_A}{3} & \frac{-d_A}{3} & \frac{\sqrt{3}}{3} \\ \frac{-d_A}{3} & \frac{-d_A}{3} & \frac{\sqrt{3}}{3} \\ \frac{-d_A}{3} & \frac{2d_A}{3} & \frac{\sqrt{3}}{3} \end{bmatrix}$$

If we now recall relation (1.82), the above determinant simplifies to

$$\frac{2}{3}d_A^2 \sin\phi = \frac{1}{3^3}\det\begin{bmatrix} 2d_A & -d_A & \sqrt{3} \\ -d_A & -d_A & \sqrt{3} \\ -d_A & 2d_A & \sqrt{3} \end{bmatrix}$$

Upon expansion of the determinant by cofactors of its third column, we obtain

$$\begin{aligned} \frac{2}{3}d_A^2 \sin\phi &= \frac{\sqrt{3}}{3^3}\left[\det\begin{bmatrix} -d_A & -d_A \\ -d_A & 2d_A \end{bmatrix} - \det\begin{bmatrix} 2d_A & -d_A \\ -d_A & 2d_A \end{bmatrix} + \det\begin{bmatrix} 2d_A & -d_A \\ -d_A & -d_A \end{bmatrix}\right] \\ &= \frac{\sqrt{3}}{3^3}[(-2d_A^2 - d_A^2) - (4d_A^2 - d_A^2) + (-2d_A^2 - d_A^2)] \\ &= \frac{\sqrt{3}}{3^3}(-9d_A^2) = -\frac{\sqrt{3}}{3}d_A^2 \end{aligned}$$

and hence

$$\sin\phi = -\frac{\sqrt{3}}{2}$$

which verifies $\sin^2\phi + \cos^2\phi = 1$ with the result obtained above for $\cos\phi$. We can thus conclusively state that, for the above value of $\boldsymbol{e}$, $\phi = 240°$. Of course, a reversal of the sign of the components of $\boldsymbol{e}$ will lead to the value $\phi = 120°$. One or the other defines $\mathcal{L}$.

4.3.4 Reflection

A reflection, similar to a rotation, preserves the distance between every two points of an object, but changes its *hand*. For example, making abstraction of the internal organs—and of the hairstyle, of course—the human body can be regarded as a symmetric object, its plane of symmetry being the *sagittal plane*. This plane divides the body into two symmetric parts, left and right. The left part is a *reflection* of the right part, the sagittal plane thus being the *plane of reflection*. As a matter of fact, the plane qualifier derives from Latin "sagitta," which means "arrow".[8]

A three-dimensional reflection (mirroring) is usually obtained by coordinate transformations about a specified reflection plane.

- The matrix representing a reflection about the plane $x = 0$ is given by

$$\boldsymbol{M}_X = \begin{bmatrix} -1 & 0 & 0 \\ 0 & 1 & 0 \\ 0 & 0 & 1 \end{bmatrix} \tag{4.41}$$

[8]The qualifier must come from the plane containing the arrow upon being thrown by the archer.

- The matrix representing a reflection about the plane $y = 0$ is given, in turn, by

$$\boldsymbol{M}_Y = \begin{bmatrix} 1 & 0 & 0 \\ 0 & -1 & 0 \\ 0 & 0 & 1 \end{bmatrix} \tag{4.42}$$

- The matrix representing a reflection about the plane $z = 0$ is, finally, given by

$$\boldsymbol{M}_Z = \begin{bmatrix} 1 & 0 & 0 \\ 0 & 1 & 0 \\ 0 & 0 & -1 \end{bmatrix} \tag{4.43}$$

To reflect an object about any arbitrary plane, combined transformations involving rotations and reflections will have to be produced. In the special case in which the reflection plane, call it Π_r, passes through the origin, with *unit* normal $\boldsymbol{n}$, the reflection matrix $\boldsymbol{M}$ takes the simple form

$$\boldsymbol{M} = \mathbf{1} - 2\boldsymbol{n}\boldsymbol{n}^T \tag{4.44}$$

The reader is invited to prove that any point of position vector $\boldsymbol{p}$ is reflected by $\boldsymbol{M}$ into a vector $\boldsymbol{r}$ on the opposite side of Π_r.

4.4 The Most General Affine Transformation in 3D

The most general affine transformation in 3D involves, as does its counterpart in 2D, studied in Sect. 4.2, a combination of scalings, translations, rotations, and reflections. We study here these transformations, as pertaining to 3D objects, by considering, again, two cases: (i) the transformation leaves the origin fixed and (ii) the transformation carries the origin O to a new position, O'.

Case (i): Given that no translation of the origin is involved, the problem reduces to finding matrix $\boldsymbol{M}$ of the homogeneous transformation, and hence we have to find *nine unknowns*, i.e., its nine distinct entries. We thus need nine equations relating the two objects, which can be derived from the correspondence between three points in $\mathcal{O}$ and three points in $\mathcal{O}'$, the original and the final locations of the object. Let these points be P_1, P_2, and P_3 in $\mathcal{O}$ and their corresponding *images*, P_1', P_2', and P_3' in $\mathcal{O}'$. As each point correspondence entails three scalar equations, one for each coordinate, we have a total of nine scalar equations, enough to compute the nine unknowns. Let, moreover, $\boldsymbol{p}_i$ and $\boldsymbol{p}_i'$ be the position vector of P_i and P_i'. Hence,

$$\boldsymbol{p}_1' = \boldsymbol{M}\boldsymbol{p}_1, \quad \boldsymbol{p}_2' = \boldsymbol{M}\boldsymbol{p}_2, \quad \boldsymbol{p}_3' = \boldsymbol{M}\boldsymbol{p}_3 \tag{4.45a}$$

Now we introduce two 3×3 matrix arrays:

$$\boldsymbol{P} \equiv \begin{bmatrix} \boldsymbol{p}_1 & \boldsymbol{p}_2 & \boldsymbol{p}_3 \end{bmatrix}, \quad \boldsymbol{P}' \equiv \begin{bmatrix} \boldsymbol{p}'_1 & \boldsymbol{p}'_2 & \boldsymbol{p}'_3 \end{bmatrix} \tag{4.45b}$$

In this way, the three vector equations (4.45a) can be cast in the form of one single 3×3 *matrix equation*:

$$\boldsymbol{P}' = \boldsymbol{M}\boldsymbol{P} \tag{4.46}$$

from which we can solve for $\boldsymbol{M}$ by multiplying both sides by $\boldsymbol{P}^{-1}$ from the right, *provided that* $\boldsymbol{P}$ *is not singular*. This will be the case *if (i) the plane of the three points chosen,* P_1, P_2, *and* P_3, *does not include the origin and (ii) the points are not aligned.*[9] With this provision, then, Eq. (4.46) can be solved for $\boldsymbol{M}$ in the form

$$\boldsymbol{M} = \boldsymbol{P}'\boldsymbol{P}^{-1} \tag{4.47}$$

thereby completing the solution.

Case (ii): In this case, we have 12 unknowns, the nine entries of $\boldsymbol{M}$, as in Case (i), plus the three components of vector $\boldsymbol{t}$. We thus need 12 scalar equations to find the foregoing unknowns. These equations are readily obtained by establishing the correspondence between two quadruplets of points, one in $\mathcal{O}$ and one in $\mathcal{O}'$. This correspondence should be established between homogeneous coordinates, as $\boldsymbol{t}$ is not included in $\boldsymbol{M}$. We proceed as in Case (i): let now $\boldsymbol{p}_i$ and $\boldsymbol{p}'_i$ be the arrays of *homogeneous coordinates* of P_i and P'_i, respectively, for $i = 1, 2, 3, 4$, and $\boldsymbol{T}$ the 4×4 homogeneous-transformation matrix sought. The four correspondences then follow:

$$\boldsymbol{p}'_i = \boldsymbol{T}\boldsymbol{p}_i, \quad \text{for} \quad i = 1, 2, 3, 4 \tag{4.48}$$

By defining the 4×4 arrays $\boldsymbol{P}$ and $\boldsymbol{P}'$ as

$$\boldsymbol{P} = \begin{bmatrix} \boldsymbol{p}_1 & \boldsymbol{p}_2 & \boldsymbol{p}_3 & \boldsymbol{p}_4 \end{bmatrix}, \quad \boldsymbol{P}' = \begin{bmatrix} \boldsymbol{p}'_1 & \boldsymbol{p}'_2 & \boldsymbol{p}'_3 & \boldsymbol{p}'_4 \end{bmatrix} \tag{4.49}$$

the four correspondences (4.48) can now be cast in the form of a 4×4 *matrix equation*:

$$\boldsymbol{P}' = \boldsymbol{T}\boldsymbol{P} \tag{4.50}$$

from which we can solve for $\boldsymbol{T}$ upon multiplying both sides of the foregoing equation by $\boldsymbol{P}^{-1}$ from the right, as long, of course, as $\boldsymbol{P}$ is not singular. Notice that, although homogeneous transformations representing affine transformations are never singular, matrices $\boldsymbol{P}$ and $\boldsymbol{P}'$ can become so. As the reader is invited to verify, in Exercise 4.2, these matrices become singular if and only if the four given points are coplanar. If this is not the case, then Eq. (4.50) can be solved for $\boldsymbol{T}$ in the form

$$\boldsymbol{T} = \boldsymbol{P}'\boldsymbol{P}^{-1} \tag{4.51}$$

[9] See Exercise 4.1.

thereby completing the solution of the second case.

4.4.1 Examples

Example 4.4.1 Find the affine transformation that carries the unit sphere centered at the origin of the 3D space into the ellipsoid of semiaxes 0.5, 2, and 3, centered at the origin as well, with the ellipsoid axes defined by the unit vectors $\boldsymbol{e}_i$, for $i = 1, 2, 3$, given below. The axes are numbered from shortest to longest, their corresponding unit vectors being

$$\boldsymbol{e}_1 = \begin{bmatrix} -1/3 \\ 2/3 \\ 2/3 \end{bmatrix}, \quad \boldsymbol{e}_2 = \begin{bmatrix} 2/3 \\ -1/3 \\ 2/3 \end{bmatrix}, \quad \boldsymbol{e}_3 = \begin{bmatrix} 2/3 \\ 2/3 \\ -1/3 \end{bmatrix}$$

Find also the inverse transformation that carries the ellipsoid back into the original unit sphere.

Solution Three points A, B, C are chosen on the sphere, at the intersections of the sphere with the coordinate axes, and their corresponding images, A', B', C', on the ellipsoid, the position vectors of these points being labeled $\boldsymbol{a}$, $\boldsymbol{b}$, ..., $\boldsymbol{c}'$. These are listed below:

$$\boldsymbol{a} = \begin{bmatrix} 1 \\ 0 \\ 0 \end{bmatrix}, \quad \boldsymbol{b} = \begin{bmatrix} 0 \\ 1 \\ 0 \end{bmatrix}, \quad \boldsymbol{c} = \begin{bmatrix} 0 \\ 0 \\ 1 \end{bmatrix}$$

$$\boldsymbol{a}' = l_1\boldsymbol{e}_1 = \begin{bmatrix} -1/6 \\ 1/3 \\ 1/3 \end{bmatrix}, \quad \boldsymbol{b}' = l_2\boldsymbol{e}_2 = \begin{bmatrix} 4/3 \\ -2/3 \\ 4/3 \end{bmatrix}, \quad \boldsymbol{c}' = l_3\boldsymbol{e}_3 = \begin{bmatrix} 2 \\ 2 \\ -1 \end{bmatrix}$$

where l_i denotes the length of the ith semiaxes, matrices $\boldsymbol{P}$ and $\boldsymbol{P}'$ thus being readily constructed:

$$\boldsymbol{P} = \mathbf{1}, \quad \boldsymbol{P}' = \begin{bmatrix} -1/6 & 4/3 & 2 \\ 1/3 & -2/3 & 2 \\ 1/3 & 4/3 & -1 \end{bmatrix}$$

with $\mathbf{1}$ representing the 3×3 identity matrix. The matrix $\boldsymbol{M}$ sought is thus obtained as

$$\boldsymbol{M} = \boldsymbol{P}'\boldsymbol{P}^{-1} = \boldsymbol{P}' = \begin{bmatrix} -1/6 & 4/3 & 2 \\ 1/3 & -2/3 & 2 \\ 1/3 & 4/3 & -1 \end{bmatrix}$$

thereby completing the calculation of $\boldsymbol{M}$, its inverse being computed by means of expression (1.118)

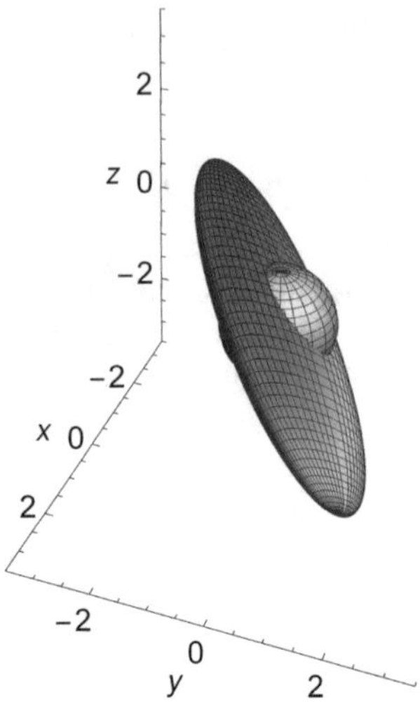

Fig. 4.22 An affine transformation of the unit sphere centered at the origin into an ellipsoid of semiaxes 0.5, 2, and 3, centered likewise

$$\boldsymbol{M}^{-1} = \begin{bmatrix} -1/6 & 4/3 & 4/3 \\ 1/3 & -1/6 & 1/3 \\ 2/9 & 2/9 & -1/9 \end{bmatrix} \tag{4.52}$$

Shown in Fig. 4.22 is a rendering of the two surfaces, with the two sharing the same center.

Example 4.4.2 We revisit here Example 4.4.1, but now the center of the ellipsoid is to be located at point $O'(3,\ 2,\ 1)$. Find both the 4×4 homogeneous-transformation matrix representing the underlying affine transformation and its inverse.

Solution Three points A, B, C and their images, A', B', C', are chosen as in Example 4.4.1. One fourth point is needed now, along with its image. The obvious choice is the center of the sphere, i.e., the origin O, its image being the center of the ellipsoid, O'. The position vectors of the first three points are exactly the same as in Example 4.4.1, the fourth point being the origin O of position vector $\boldsymbol{o}$; the images of the first three points are those of the foregoing example, if augmented by the translation $\boldsymbol{t} = [3,\ 2,\ 1]^T$. Obviously, the position vectors of O and O' are

$$\boldsymbol{o} = \boldsymbol{0}, \quad \boldsymbol{o}' = \begin{bmatrix} 3 \\ 2 \\ 1 \end{bmatrix}$$

The 4×4 matrices $\boldsymbol{P}$ and $\boldsymbol{P}'$ are readily constructed:

$$P = \begin{bmatrix} 1 & 0 & 0 & 0 \\ 0 & 1 & 0 & 0 \\ 0 & 0 & 1 & 0 \\ 1 & 1 & 1 & 1 \end{bmatrix}, \quad P' = \begin{bmatrix} 17/6 & 13/3 & 5 & 3 \\ 7/3 & 4/3 & 4 & 2 \\ 4/3 & 7/3 & 0 & 1 \\ 1 & 1 & 1 & 1 \end{bmatrix}$$

The matrix $\boldsymbol{T}$ sought is thus obtained from

$$\boldsymbol{T} = \boldsymbol{P}'\boldsymbol{P}^{-1}$$

where $\boldsymbol{P}^{-1}$ is yet to be computed. This is done most safely by recalling the expression for a block matrix, of Eq. (1.86). Indeed, given its simple pattern, $\boldsymbol{P}$ can be readily partitioned, namely,

$$\boldsymbol{P} = \begin{bmatrix} \mathbf{1} & \mathbf{0} \\ \boldsymbol{e}^T & 1 \end{bmatrix}, \quad \boldsymbol{e} \equiv \begin{bmatrix} 1 \\ 1 \\ 1 \end{bmatrix}$$

with $\mathbf{1}$ denoting the 3×3 identity matrix. The inverse of the above matrix can thus be readily obtained from its block form:

$$\boldsymbol{P}^{-1} = \begin{bmatrix} \mathbf{1} & \mathbf{0} \\ -\boldsymbol{e}^T & 1 \end{bmatrix} = \begin{bmatrix} 1 & 0 & 0 & 0 \\ 0 & 1 & 0 & 0 \\ 0 & 0 & 1 & 0 \\ -1 & -1 & -1 & 1 \end{bmatrix}$$

Hence, the desired 4×4 transformation matrix $\boldsymbol{T}$ is

$$\boldsymbol{T} = \begin{bmatrix} -1/6 & 4/3 & 2 & 3 \\ 1/3 & -2/3 & 2 & 2 \\ 1/3 & 4/3 & -1 & 1 \\ 0 & 0 & 0 & 1 \end{bmatrix}$$

thereby completing the calculation of $\boldsymbol{T}$, its inverse being

$$\boldsymbol{T}^{-1} = \begin{bmatrix} -2/3 & 4/3 & 4/3 & -2 \\ 1/3 & -1/6 & 1/3 & -1 \\ 2/9 & 2/9 & -1/9 & -1 \\ 0 & 0 & 0 & 1 \end{bmatrix} \tag{4.53}$$

Shown in Fig. 4.23 is a rendering of the two surfaces, showing the offset between the two centers.

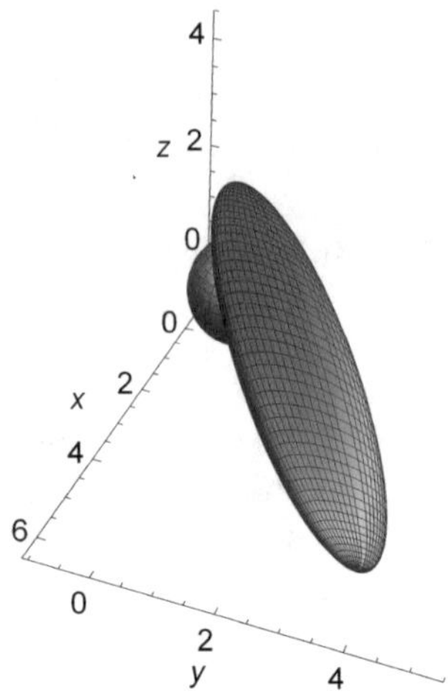

Fig. 4.23 An affine transformation of the unit sphere centered at the origin into an ellipsoid of semiaxes 0.5, 2, and 3, centered offset with respect to the sphere

4.4.2 Affine Transformations of 3D Objects Under Implicit Representations

The application of affine transformations to create complex shapes from simple ones, as outlined in the preamble of this section, is limited to parametric representations of the surfaces involved. When such representations are either not immediately available or too limited to represent the whole surface at hand, then implicit representations should be used. Similar to the case of affine transformations in 2D, the implementation of affine transformations in 3D calls for multiplying the position vector $\boldsymbol{p}$ in homogeneous coordinates by the *inverse* of the desired transformation, thereby obtaining a new vector $\boldsymbol{q}$ in the form

$$\boldsymbol{q} = \boldsymbol{T}^{-1}\boldsymbol{p} \tag{4.54}$$

We illustrate this concept with examples.

Example 4.4.3 We revisit here Example 4.4.2. Using an implicit representation of the unit sphere, find the affine transformation required to map the sphere into the desired three-axis ellipsoid. Also find the implicit representation of the ellipsoid.

Solution The affine transformation required is a composition of: (a) a scaling $\boldsymbol{S}$ of factors $S_x = 0.5$, $S_y = 2$, and $S_z = 3$; (b) a translation $\boldsymbol{t} = [3,\ 2,\ 1]^T$ that will take the center of the sphere O into the center of the ellipsoid O'; and (c) a rotation that will rotate the sphere from an attitude in which the three intersection points of the sphere with the coordinate axes—those that define the first *octant*[10]—lie in the corresponding axes of the ellipsoid. The matrix $\boldsymbol{R}$ representing this rotation can be produced by means of the procedure outlined in Sect. 4.4 to obtain matrix $\boldsymbol{M}$ of the homogeneous transformation in the absence of a shift of the origin. We recall below the technique used in the above section to produce the desired rotation matrix.

Regard the position vectors of three points on the unit sphere as the three unit vectors $\boldsymbol{i}, \boldsymbol{j}, \boldsymbol{k}$ parallel to the three coordinate axes, their images being the three unit vectors along the axes of the ellipsoid, and labeled $\boldsymbol{e}_1, \boldsymbol{e}_2, \boldsymbol{e}_3$. Matrices $\boldsymbol{P}$ and $\boldsymbol{P}'$ of Sect. 4.4, Case i, are thus

$$\boldsymbol{P} = \begin{bmatrix} \boldsymbol{i} & \boldsymbol{j} & \boldsymbol{k} \end{bmatrix} = \mathbf{1} = \boldsymbol{P}^{-1} \tag{4.55a}$$

$$\boldsymbol{P}' = \begin{bmatrix} \boldsymbol{e}_1 & \boldsymbol{e}_2 & \boldsymbol{e}_3 \end{bmatrix} = \begin{bmatrix} -1/3 & 2/3 & 2/3 \\ 2/3 & -1/3 & 2/3 \\ 2/3 & 2/3 & -1/3 \end{bmatrix} \tag{4.55b}$$

where $\mathbf{1}$ is the 3×3 identity matrix. Matrix $\boldsymbol{R}$ is, then,

$$\boldsymbol{R} = \boldsymbol{P}'\boldsymbol{P}^{-1} = \begin{bmatrix} -1/3 & 2/3 & 2/3 \\ 2/3 & -1/3 & 2/3 \\ 2/3 & 2/3 & -1/3 \end{bmatrix}$$

from which we can now produce the 4×4 homogeneous-transformation matrices $\boldsymbol{T}_s$, $\boldsymbol{T}_t$, and $\boldsymbol{T}_r$ representing the scaling, translation, and the rotation, respectively,

$$\boldsymbol{T}_s = \begin{bmatrix} 1/2 & 0 & 0 & 0 \\ 0 & 2 & 0 & 0 \\ 0 & 0 & 3 & 0 \\ 0 & 0 & 0 & 1 \end{bmatrix}, \quad \boldsymbol{T}_t = \begin{bmatrix} 1 & 0 & 0 & 3 \\ 0 & 1 & 0 & 2 \\ 0 & 0 & 1 & 1 \\ 0 & 0 & 0 & 1 \end{bmatrix}, \quad \boldsymbol{T}_r = \begin{bmatrix} -1/3 & 2/3 & 2/3 & 0 \\ 2/3 & -1/3 & 2/3 & 0 \\ 2/3 & 2/3 & -1/3 & 0 \\ 0 & 0 & 0 & 1 \end{bmatrix}$$

Further, the equivalent transformation $\boldsymbol{T}$ is obtained, upon application of the scaling first, then the translation, and finally the rotation:

$$\boldsymbol{T} = \boldsymbol{T}_r \boldsymbol{T}_t \boldsymbol{T}_s = \begin{bmatrix} -1/6 & 4/3 & 2 & 1 \\ 1/3 & -2/3 & 2 & 2 \\ 1/3 & 4/3 & -1 & 3 \\ 0 & 0 & 0 & 1 \end{bmatrix} \tag{4.56}$$

However, for the rendering of the ellipse, we need to multiply the vector $\boldsymbol{p}$ of homogeneous coordinates of a point P on the sphere by $\boldsymbol{T}^{-1}$. We thus calculate the latter

---

[10]In the same way as the coordinate axes X and Y divide the plane into four *quadrants*, in 3D space the coordinate axes divide the space into eight *octants*.

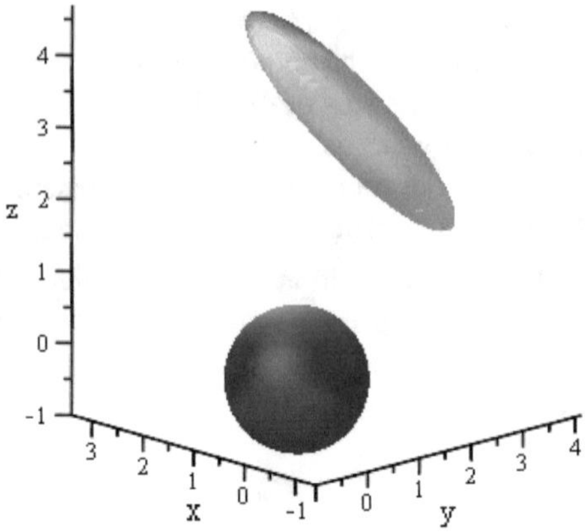

Fig. 4.24 The unit sphere centered at the origin and its distorted, displaced image after the affine transformation of Eq. (4.56)

using the general form of the inverse of a 4×4 homogeneous-transformation matrix, Eq. (1.121), both $\boldsymbol{T}^{-1}$ and $\boldsymbol{p}$ being displayed below:

$$\boldsymbol{T}^{-1} = \begin{bmatrix} -2/3 & 4/3 & 4/3 & -6 \\ 1/3 & -1/6 & 1/3 & -1 \\ 2/9 & 2/9 & -1/9 & -1/3 \\ 0 & 0 & 0 & 1 \end{bmatrix}, \quad \boldsymbol{p} = \begin{bmatrix} x \\ y \\ z \\ 1 \end{bmatrix}$$

Hence, the image $\boldsymbol{q}$ of vector $\boldsymbol{p}$ becomes

$$\boldsymbol{q} = \boldsymbol{T}^{-1}\boldsymbol{p} = \begin{bmatrix} -2x/3 + 4y/3 + 4z/3 - 6 \\ x/3 - y/6 + z/3 - 1 \\ 2x/9 + 2y/9 - z/9 - 1/3 \\ 1 \end{bmatrix}$$

whose components, when substituted into the equation of the unit sphere, yield the implicit equation of the ellipsoid, namely,

$$\begin{aligned} \mathcal{E}:\ & q_1^2 + q_2^2 + q_3^2 - 1 \\ \equiv & \frac{194}{27}x - \frac{448}{27}z - \frac{145}{81}xy - \frac{130}{81}xz + \frac{275}{81}yz - \frac{427}{27}y \\ & + \frac{49}{81}x^2 + \frac{601}{324}y^2 + \frac{154}{81}z^2 + \frac{325}{9} = 0 \end{aligned}$$

Sphere and ellipsoid are displayed in Fig. 4.24.

4.5 Techniques for 3D Object Modeling

Three-dimensional objects can be regarded as regions of space bounded by closed surfaces. In this section, we study the various techniques available for the production of such surfaces.

4.5.1 Surfaces of Revolution

A simple family of surfaces is obtained by rotating a planar curve around an axis in the plane of the curve, thereby obtaining a surface of revolution. For example, a circular cylinder is formed by rotating a line segment parallel to the Z-axis through an angle of 2π around the same Z-axis.

We will describe here the generation of surfaces of revolution by means of the rotation of a planar curve Γ in the XZ-plane, the *generatrix*, about the Z-axis. Shown in Fig. 4.25 is the generatrix Γ and the displacement of one arbitrary point of Γ upon a rotation of Γ about Z through an angle θ.

The homogeneous coordinates of P and P' are stored in the four-dimensional arrays $\boldsymbol{p}$ and $\boldsymbol{p}'$ which are related by an affine transformation of the form of Eq. (4.30a), with $\boldsymbol{M}$ representing a rotation about the Z-axis through an angle θ, namely,

$$\boldsymbol{M} = \begin{bmatrix} \cos\theta & -\sin\theta & 0 \\ \sin\theta & \cos\theta & 0 \\ 0 & 0 & 1 \end{bmatrix}$$

For the case depicted in Fig. 4.25, we have, in homogeneous-coordinate form,

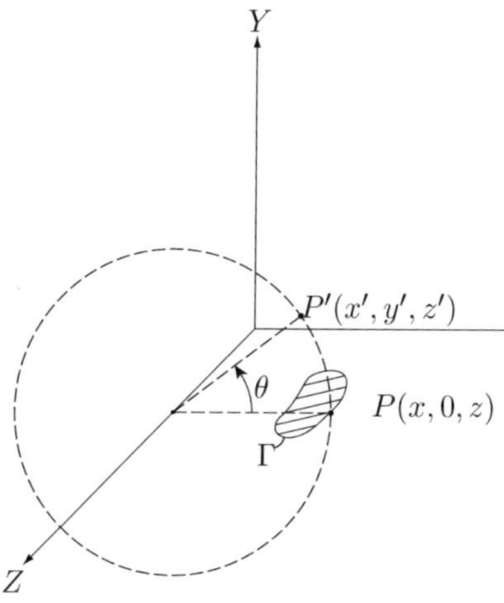

Fig. 4.25 Generation of a surface of revolution by means of the rotation of a generatrix Γ in the XZ-plane about the Z-axis

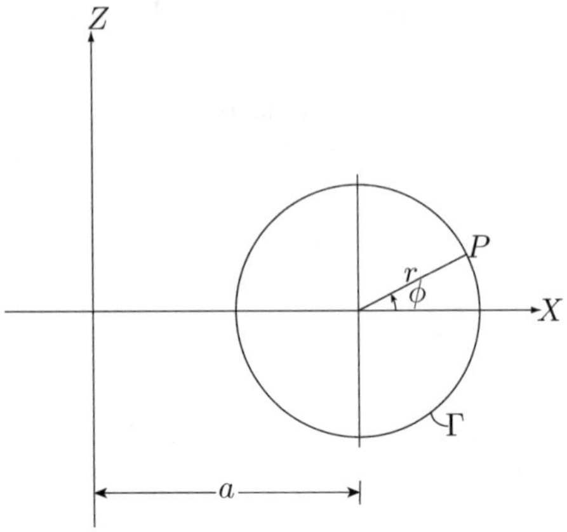

Fig. 4.26 Construction of an O-ring by revolving the circle Γ about the Z-axis

$$\boldsymbol{p} = \begin{bmatrix} x \\ 0 \\ z \\ 1 \end{bmatrix}, \quad \boldsymbol{p}' = \begin{bmatrix} \cos\theta & -\sin\theta & 0 & 0 \\ \sin\theta & \cos\theta & 0 & 0 \\ 0 & 0 & 1 & 0 \\ 0 & 0 & 0 & 1 \end{bmatrix} \begin{bmatrix} x \\ 0 \\ z \\ 1 \end{bmatrix}$$

i.e.,

$$\boldsymbol{p}' = \begin{bmatrix} x\cos\theta \\ x\sin\theta \\ z \\ 1 \end{bmatrix}$$

Example 4.5.1 Construct an O-ring of cross section of radius r and main radius $a > r$.

Solution The O-ring is generated by revolving the circle Γ lying in the XZ-plane, as depicted in Fig. 4.26, about the Z-axis. Representing the circle in polar coordinates, we have

$$\boldsymbol{p} = \begin{bmatrix} a + r\cos\phi \\ 0 \\ r\sin\phi \\ 1 \end{bmatrix} \Rightarrow \boldsymbol{p}' = \begin{bmatrix} (a + r\cos\phi)\cos\theta \\ (a + r\cos\phi)\sin\theta \\ r\sin\phi \\ 1 \end{bmatrix}$$

A small piece of code was written using computer-algebra software[11] to produce a rendering of the O-ring, namely,

[11] Maplesoft's Maple 10.

```
> restart; with(plots):
> with(LinearAlgebra):
> R:=«cos(theta),sin(theta),0,0>|<sin(theta),cos
> (theta),0,0>|
> <0,0,1,0>|<0,0,0,1»;
```

$$\begin{bmatrix} \cos(\theta) & \sin(\theta) & 0 & 0 \\ \sin(\theta) & \cos(\theta) & 0 & 0 \\ 0 & 0 & 1 & 0 \\ 0 & 0 & 0 & 1 \end{bmatrix}$$

```
> p:=<a+r*cos(phi),0,r*sin(phi),1>;
```

$$\begin{bmatrix} 3+\cos(\phi) \\ 0 \\ \sin(\phi) \\ 1 \end{bmatrix}$$

```
> Rp:=R.p;
```

$$\begin{bmatrix} \cos(\theta)(3+\cos(\phi)) \\ \sin(\theta)(3+\cos(\phi)) \\ \sin(\phi) \\ 1 \end{bmatrix}$$

```
> a:=3; r:=1;
```

$$3$$
$$1$$

```
> plot3d(Rp[1..3], theta=0..2*Pi,
> phi=0..2*Pi, scaling=constrained,grid=[60,60]);
```

The code produced the rendering of Fig. 4.27, with the numerical values $r = 10\,\text{mm}$, $a = 50\,\text{mm}$.

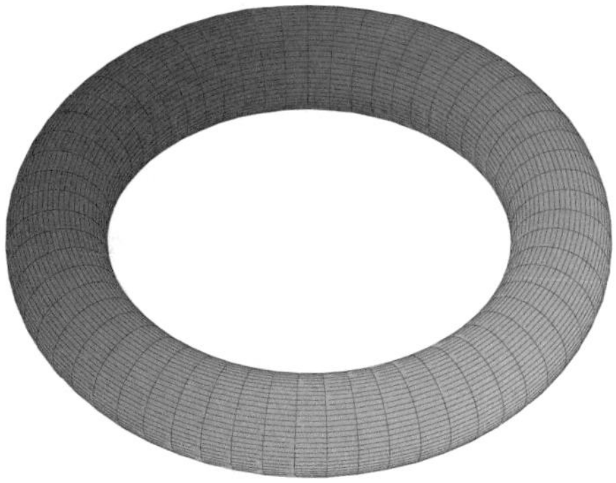

Fig. 4.27 Computer rendering of an O-ring with $r = 10\,\text{mm}$, $a = 50\,\text{mm}$

4.5.2 Extrusion

Extrusion is a procedure by which a surface is *generated* through the movement of a line segment, a curve segment, a polygon, and so forth, i.e., a *generatrix*, along a *defined path*. The generatrix of an extrusion operation can be a straight line or a curve. The corresponding extruded surface is represented in parametric form as follows:

$$\boldsymbol{q}(t, s) = \boldsymbol{T}(s)\boldsymbol{p}(t) \tag{4.57}$$

where $\boldsymbol{p}(t)$ is the four-dimensional array of homogeneous coordinates of a point P of the generatrix, in parametric form, and $\boldsymbol{T}(s)$ is the extrusion transformation based on the shape of the path, given in terms of a second parameter, s, and $\boldsymbol{q}(t, s)$ is the four-dimensional array of homogeneous coordinates of the transformed point Q.

The extrusion transformation can contain translations, scalings, rotations, or combinations of these transformations. For the case in which the path is a line, all the points of the generatrix Γ, which we will assume to be a planar curve in the XZ-plane, translate in the direction of extrusion, given by a unit vector $\boldsymbol{e}$.

The displacement of every point P of Γ is thus $s\boldsymbol{e}$, where $s \geq 0$ is the translation parameter, the *extrusion matrix* then taking the form

$$\boldsymbol{T}(s) = \begin{bmatrix} \boldsymbol{M} & s\boldsymbol{e} \\ \boldsymbol{0}^T & 1 \end{bmatrix}$$

Matrix $\boldsymbol{M}$ can be constant or a function of s, depending on the type of extrusion at hand. Moreover, while we have assumed that Γ lies in the XZ-plane, we need not impose any constraint on $\boldsymbol{e}$, except that it is a unit vector. A few examples will illustrate the power of the extrusion transformation to construct a variety of solids.

Example 4.5.2 A shaft of radius r and length l can be constructed by the *simple extrusion* of a circle centered at the origin of the XZ-plane, with direction of extrusion given by the Y-axis. In this case,

$$\boldsymbol{p}(t) = \begin{bmatrix} r\cos t \\ 0 \\ r\sin t \\ 1 \end{bmatrix}, \quad \boldsymbol{e} = \begin{bmatrix} 0 \\ 1 \\ 0 \end{bmatrix}, \quad \boldsymbol{M} = \boldsymbol{1}, \quad 0 \leq s \leq l$$

Hence,

$$\boldsymbol{q}(t, s) = \begin{bmatrix} 1 & 0 & 0 & 0 \\ 0 & 1 & 0 & s \\ 0 & 0 & 1 & 0 \\ 0 & 0 & 0 & 1 \end{bmatrix} \begin{bmatrix} r\cos t \\ 0 \\ r\sin t \\ 1 \end{bmatrix} = \begin{bmatrix} r\cos t \\ s \\ r\sin t \\ 1 \end{bmatrix}$$

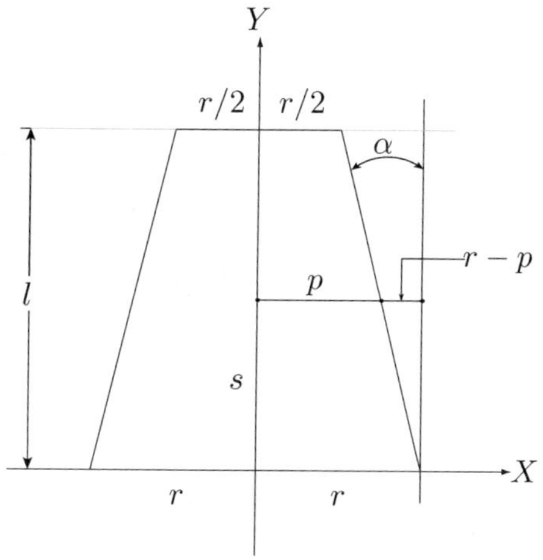

Fig. 4.28 Construction of a tapered shaft

Example 4.5.3 A *tapered shaft* of largest radius r, smallest radius $r/2$, length l, and tapering angle α can be constructed using the same generatrix Γ and the same direction of extrusion as in Example 4.5.2. The difference now is that $\boldsymbol{M}$ involves a scaling by the angle of tapering, so that

$$\boldsymbol{M} = k(s)\mathbf{1}$$

where $k(s)$ is a scaling factor, which is determined with the aid of Fig. 4.28. Notice that α can be computed from the dimensions r and l.

The reader should be able to verify that the scaling factor $k(s)$ is given by

$$k(s) \equiv \frac{p}{r} = 1 - \frac{s}{2l}$$

A three-dimensional rendering of the shaft, for $l = 300$ mm and $r = 60$ mm, is included in Fig. 4.29.

Example 4.5.4 Construction of a screw. We show in Fig. 4.30a the sketch of a common type of screw, illustrating its terminology. In this sketch, the vee-shaped crests and roots serve purposes of sketch simplicity; in practice, these are either flattened, as illustrated in Fig. 4.30b, or rounded. Devise a means of constructing the threaded portion of the screw. For simplicity, keep the vee-shapes.

Solution We shall use *extrusion* along the direction of the axis $\mathcal{L}$ of the screw combined with rotation about the same axis, the generatrix Γ being illustrated in Fig. 4.31a.

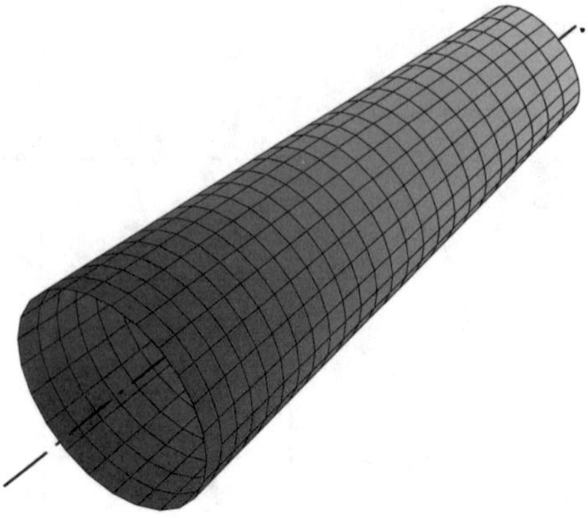

Fig. 4.29 Three-dimensional rendering (not a perspective) of the tapered shaft, with $l = 300\,\text{mm}$ and $r = 60\,\text{mm}$

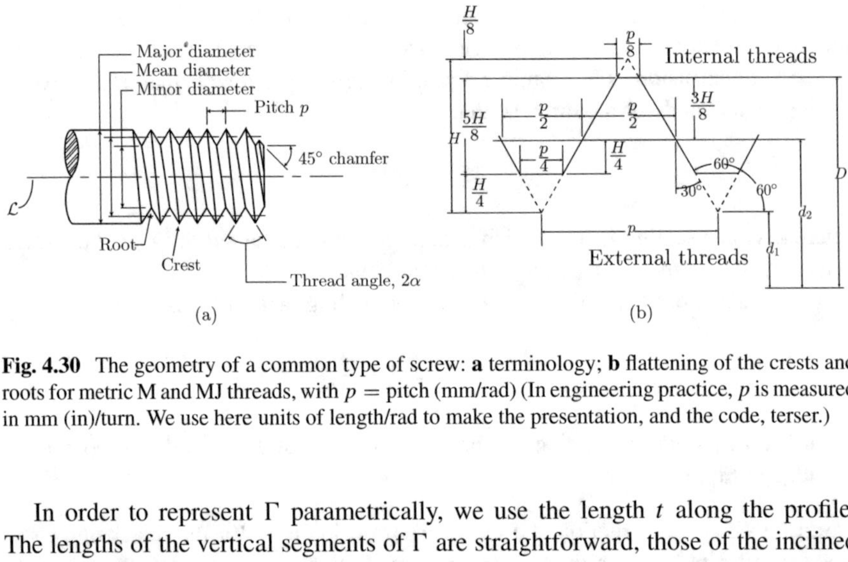

Fig. 4.30 The geometry of a common type of screw: **a** terminology; **b** flattening of the crests and roots for metric M and MJ threads, with $p =$ pitch (mm/rad) (In engineering practice, p is measured in mm (in)/turn. We use here units of length/rad to make the presentation, and the code, terser.)

In order to represent Γ parametrically, we use the length t along the profile. The lengths of the vertical segments of Γ are straightforward, those of the inclined segments, of length l, being derived from the detail in Fig. 4.31b: Because l and p are sides of the same equilateral triangle,

$$l = p$$

Hence, Γ is described, parametrically, in *piecewise form* in terms of the parameter t, as

$$l = p$$

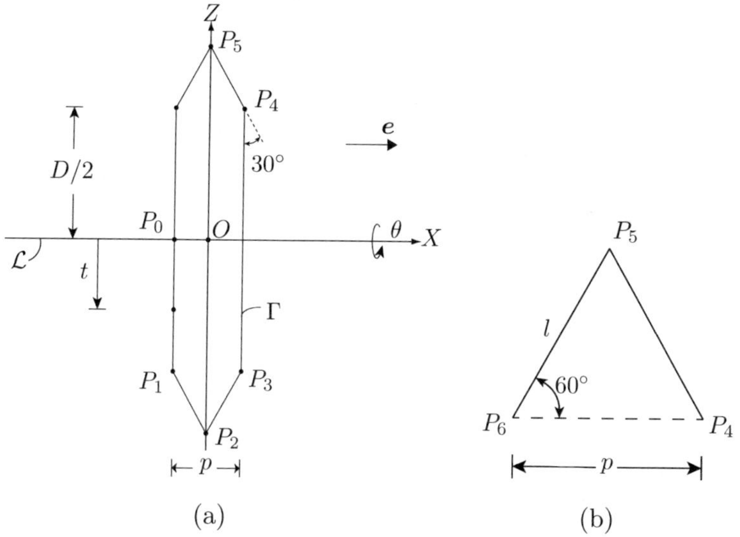

Fig. 4.31 The generatrix Γ for the construction of the threaded surface of a screw: **a** general layout; **b** detail of the vee-shaped parts

$$
\begin{array}{lll}
0 \le t \le \dfrac{D}{2}: & x = -\dfrac{p}{2}, & z = -t \\
\dfrac{D}{2} \le t \le \dfrac{D}{2} + p: & x = -\dfrac{p}{2} + \left(t - \dfrac{D}{2}\right)\cos 60^\circ, & z = -\dfrac{D}{2} + \left(-t + \dfrac{D}{2}\right)\sin 60^\circ \\
\dfrac{D}{2} + p \le t \le \dfrac{D}{2} + 2p: & x = \left(t - \dfrac{D}{2} - p\right)\cos 60^\circ, & z = -\dfrac{D}{2} + \left(t - \dfrac{D}{2} - 2p\right)\sin 60^\circ \\
\dfrac{D}{2} + 2p \le t \le \dfrac{3D}{2} + 2p: & x = \dfrac{p}{2}, & z = t - D - 2p \\
\dfrac{3D}{2} + 2p \le t \le \dfrac{3D}{2} + 3p: & x = \dfrac{p}{2} + \left(-t + \dfrac{3D}{2} + 2p\right)\cos 60^\circ, & z = \dfrac{D}{2} + \left(t - \dfrac{3D}{2} + 2p\right)\sin 60^\circ \\
\dfrac{3D}{2} + 3p \le t \le \dfrac{3D}{2} + 4p: & x = \left(-t + \dfrac{3D}{2} + 3p\right)\cos 60^\circ, & z = \dfrac{D}{2} + \left(-t + \dfrac{3D}{2} + 4p\right)\sin 60^\circ \\
\dfrac{3D}{2} + 3p \le t \le 2D + 4p: & x = -\dfrac{p}{2}, & z = -t + 2D + 4p
\end{array}
$$

With Γ available in parametric form, all we need is $\boldsymbol{t}$ and $\boldsymbol{M}$; $\boldsymbol{t}$ is a translation in the direction of $\boldsymbol{e}$, and hence

$$\boldsymbol{t} = s\boldsymbol{e}, \quad \boldsymbol{e} = \begin{bmatrix} 1 \\ 0 \\ 0 \end{bmatrix}$$

with s measured along the X-axis, introduced in Fig. 4.31a. Moreover, $\boldsymbol{M}$ is a rotation about X of angle $\theta = s/p$; hence, recalling Eq. (4.40),

$$\boldsymbol{M} = \begin{bmatrix} 1 & 0 & 0 \\ 0 & \cos(s/p) & -\sin(s/p) \\ 0 & \sin(s/p) & \cos(s/p) \end{bmatrix}$$

The 4×4 homogeneous-transformation matrix $\boldsymbol{T}$, for the given pitch, written as a rational expression, $p = 3/2$, then becomes[12]

$$\boldsymbol{T} = \begin{bmatrix} 1 & 0 & 0 & s/3 \\ 0 & \cos(2s/3) & -\sin(2s/3) & 0 \\ 0 & \sin(2s/3) & \cos(2s/3) & 0 \\ 0 & 0 & 0 & 1 \end{bmatrix}$$

The transformation was implemented using computer algebra in the piece of code displayed below:

```
> restart: with(plots):
> with(LinearAlgebra):
> # definition of the generatrix
> Gamma
> p:=1.5: d:=10: # Notes: $p$ in mm/rad, $d$ in mm.
> Maple reserves "D" for derivative; we use "d" to represent "D" in
> the definition of the generatrix
> x1:= piecewise(t<=d/2, -p/2,
>                t<=d/2+p,-p/2+(t-d/2)*cos(Pi/3),
>                t<=d/2+2*p,(t-d/2-p)*cos(Pi/3),
>                t<=3*d/2+2*p, p/2,
>                t<=3*d/2+3*p, p/2+(-t+3*d/2+2*p)*cos(Pi/3),
>                t<=3*d/2+4*p, (-t-3*d/2+3*p)*cos(Pi/3),
>                t<=2*d+4*p,-p/2):
> z1:= piecewise(t<=d/2, -t,
>                t<=d/2+p, -d/2+(-t+d/2)*sin(Pi/3),
>                t<=d/2+2*p, -d/2+(t-d/2-2*p)*sin(Pi/3),
>                t<=3*d/2+2*p, -d+t-2*p,
>                t<=3*d/2+3*p, d/2+(t-3*d/2-2*p)*sin(Pi/3),
>                t<=3*d/2+4*p,d/2+(-t+3*d/2+4*p)*sin(Pi/3),
>                t<=2*d+4*p,-t+2*d+4*p)):
```

The piece of code below yielded the rendering shown in Fig. 4.32, with parameters $D = 10$ mm and $p = 1.5$ mm.

```
> #visualization of the generatrix
> plot([x1,z1,t=0..2*d+4*p],scaling=constrained);
> # definition of the transformation
> matrix T T:=Matrix([[1,0,0,s/(2*p)],[0,cos(s/p),-sin(s/p),0],
> [0,sin(s/p),cos(s/p),0]]);
> p:=Vector([x1,0,z1,1]):
> q:=evalm(T.p):
> # screw rendering
> plot3d([q[1],q[2],q[3]],t=0..2*d+4*p,s=0..30,scaling=constrained);
> # definitions of the translation vector
> tr:=Vector([s/(2*p),0,0]);
> # definitions of the rotation matrix
> M:=Matrix([[1,0,0],[0,cos(s/p),-sin(s/p)],[0,sin(s/p),cos(s/p)]]);
```

---

[12]Notice that $\boldsymbol{t} = (s/(2p))\boldsymbol{e}$.

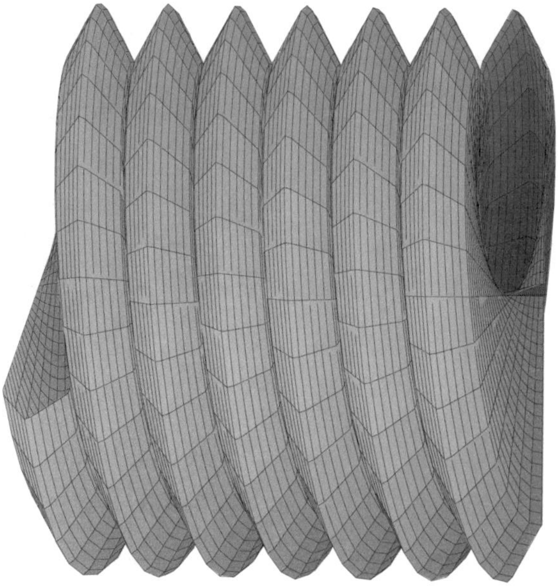

Fig. 4.32 Computer rendering of a coarse-pitch screw, with $D = 10\,\mathrm{mm}$ and $p = 1.5\,\mathrm{mm}$

```
> pp:=Vector([x1,0,z1]):
> q:=Vector(evalm(M.(pp+tr))):
> # screw rendering
> plot3d([q[1],q[2],q[3]],t=0..2*d+4*p,s=0..30,scaling=constrained,
> grid=[100,100]);
```

Once the code for producing the screw of Fig. 4.32 is available, it is now a simple matter to obtain the rendering of its left-hand counterpart. All is needed is a transformation of the vector of homogeneous coordinates $\boldsymbol{p}$ of the right-hand screw into its counterpart $\boldsymbol{q}$ of the left-hand version. This is readily done by means of a reflection $\boldsymbol{T}_{\text{ref}}$ about the X-axis, i.e., using the Y–Z-plane as the reflection plane:

$$\boldsymbol{T}_{\text{ref}} = \begin{bmatrix} -1 & 0 & 0 & 0 \\ 0 & 1 & 0 & 0 \\ 0 & 0 & 1 & 0 \\ 0 & 0 & 0 & 1 \end{bmatrix}$$

The rendering obtained with the foregoing transformations is displayed in Fig. 4.33.

4.5.2.1 Conic Extrusion

Consider a closed curve $\mathcal{G}$ lying on a sphere $\mathcal{S}_1$ of center O, as shown in Fig. 4.34. Apparently, $\mathcal{G}$ **cannot** be a planar contour, unless, of course, the contour is a circle.

Fig. 4.33 Computer rendering of a left-hand coarse-pitch screw, with $D = 10\,\mathrm{mm}$ and $p = 1.5\,\mathrm{mm}$

A *conic extrusion* is a transformation yielding a *conic surface* with generatrix $\mathcal{G}$ and vertex O. The surface is generated by $\mathcal{G}$ as the sphere $\mathcal{S}_1$ is scaled to a concentric sphere $\mathcal{S}_2$. Notice that $\mathcal{S}_2$ can be either smaller than $\mathcal{S}_1$, in which case we have an *inward extrusion*, or larger than $\mathcal{S}_1$, in which case we have an *outward extrusion*. In fact, the conic extrusion is a particular case of a 3D scaling, in which all three factors are identical, i.e., an isotropic transformation, its homogeneous-transformation matrix $\boldsymbol{T}_{ce}$ being given by

$$\boldsymbol{T}_{ce} = \begin{bmatrix} s\mathbf{1}_{3\times 3} & \mathbf{0} \\ \mathbf{0} & 1 \end{bmatrix} = \begin{bmatrix} s & 0 & 0 & 0 \\ 0 & s & 0 & 0 \\ 0 & 0 & s & 0 \\ 0 & 0 & 0 & 1 \end{bmatrix}$$

where s is the unique scaling factor.

Conic extrusions find applications in the design of spherical domes, bevel gears,[13] and spherical cam mechanisms. Shown in Fig. 4.35 is the CAD model of a pair of what is known as *conjugate spherical cams*, used in a mechanical transmission to replace bevel gears.[14] The larger cam was designed by means of a closed curve on a spherical surface $\mathcal{S}_1$, with four *lobes*, which give the cam the appearance of a four-petal flower. By means of conic extrusion, the cam was produced with a given thickness. The smaller cam was produced by means of a conic extrusion of the larger cam. In the figure, the smaller cam, moreover, was rotated about the shaft axis by an angle of 90°, i.e., 1/4 of a turn.

[13]Bevel gears were introduced in Sect. 3.3.2.

[14]Chaudhary et al. (2016).

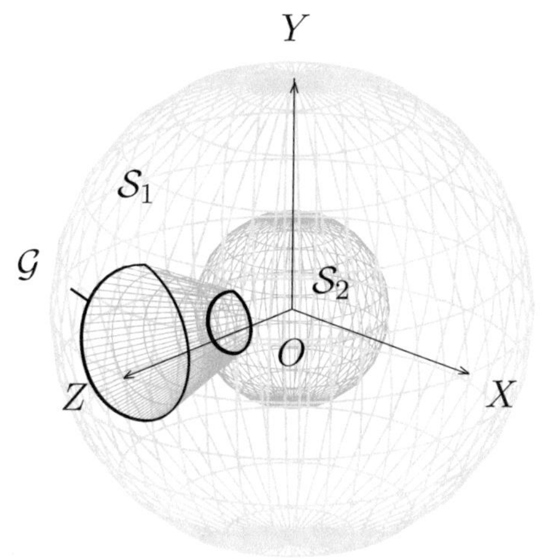

Fig. 4.34 Conic extrusion

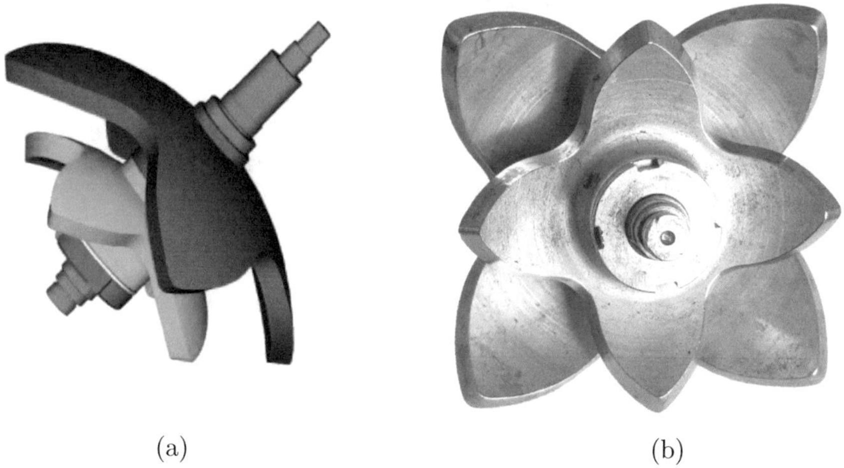

Fig. 4.35 A pair of conjugate spherical cams, obtained by means of conic extrusion: **a** computer rendering; **b** prototype (Photo courtesy of B. Belzile, Ph.D.)

4.5.3 Free-Form Surfaces

As in the case of curves, some surfaces cannot be totally described by simple formulas. Among these are surfaces used in the design of automobile bodies, ship hulls, aircraft wings, and so forth. These surfaces are usually described by a series of "patches", in the same way that a patchwork quilt is put together. The free-form curve tools, Bézier curves, B-splines, etc., can be used in free-form surface design.

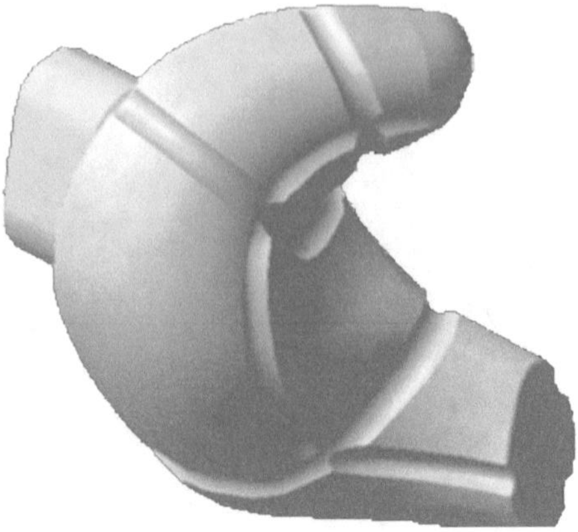

Fig. 4.36 A custom-made earplug (produced by S. Singh (2015), copyright Logitech Europe S.A., 2014. Reprinted with permission.)

These surfaces are either reproductions of natural shapes or the product of an optimization problem, such as the case of free-form curves, as discussed in Sect. 2.7. These are surfaces that do not stem from a formula of the form $f(x, y, z) = 0$. One case in point is the custom-made earplugs produced to adapt to the shape of the ear of *each individual user*. Shown in Fig. 4.36 is a CAD model produced out of a "frozen" mold of the internal ear, intended for the production of custom-made ear sleeves for plugs targeting users of carry-on audio equipment.

4.6 CAD Tools for Creating 3D Objects

Surfaces can be created using a number of different techniques supported by CAD software. The technique used is determined both by the shape to be created and the tools available in the CAD surface modeler at one's disposal. Among the most popular methods for creating surfaces, we can cite *extrusion* and *revolution*. In Sect. 4.5, we studied the homogeneous transformations involved in the construction of extruded objects and objects of revolution. Here we expand on the capabilities of CAD software for these tasks. In extrusion operations, the directrix is typically a planar curve, while the generatrix can be a line, a planar curve, or a 3D curve.

Many features in a model may be created through the use of extrusion operations. Most CAD systems use methods of automating object generation. In an extrusion operation, a closed polygon, called a profile, is drawn on a plane and is moved or swept along a defined path for a defined length. In Fig. 4.37, an example of extrusion along a line is shown. This sculpture, standing outside of Bilbao's Guggenheim Museum, was produced out of a large steel block—steel production was, until recently, the

Fig. 4.37 An example of extrusion along a line, with material removal, in art: **a** "The Gate of the Notables," a sculpture dedicated to Ramón Rubial, former Lehendakari (Head of the Provincial Government) of the Basque Country, Spain; **b** close-up of the extruded part—Art work by the Spanish Sculptor Casto Solano, 2001 (Photos courtesy of Prof. O. Altuzarra, ETSI, Bilbao, Spain)

main economic activity of Bilbao—to which material was extracted by following a profile on one of its sides.

The extruded feature will either add or subtract material from the existing model, depending on how the feature has been defined. The sculpture of Fig. 4.37 is an example of material removal in an extrusion process.

It is possible to create more complex solid models using a combination of extrusion and Boolean operations; for details on these techniques we refer the reader to the literature on engineering graphics.[15]

4.7 Summary

The book could not be complete without an invaluable tool in geometry construction, namely, *affine transformations*, which allow the designer to produce rather complex shapes out of the simpler ones studied in Chaps. 2 and 3. This chapter started with the application of such transformations to planar figures. Affine transformations, as applied not only to planar figures, but also to three-dimensional objects, are of two kinds: those that do not change the object shape, but only its position and orientation,

[15]Koenderink (1990).

i.e., its pose, and those that, besides changing the object pose, also change its shape. Of the former, we introduced affine transformations that only change the location of the object, but not its orientation; these transformations are known as *translations*. Those that change the orientation of the object, but keep one—or a set—of its points fixed are *rigid-body rotations*, or simply *rotations*. In the case of two-dimensional objects, one single point remains fixed under a rotation; in the case of their three-dimensional counterparts, a set of points remains fixed, and these lie along a line, which is the *axis of rotation*. Affine transformations that change the shape of the body allow the production of some rather complex forms starting from simple shapes. Such is the case of surfaces of a screw, as demonstrated with an example. This example is included to illustrate the power of affine transformations. The production of the screw would be extremely time-consuming if a more "direct" method of production were applied, such as obtaining, by "brute force," the curved surface via its equation. The chapter concludes with the production of *free-form surfaces*, i.e., those that arise naturally from biological organs, like the internal ear, illustrated with an example, by means of *splines*. Given the introductory nature of the book, a detailed account of splines is not included. One reference is included of a book—see footnote 16—that covers such advanced topics.

4.8 Exercises

4.1 Given three points $\{P_i\}_1^3$ in 3D space and its corresponding position vectors $\{\boldsymbol{p}_i\}_1^3$, show that:

(a) if the plane of the three points includes the origin, then the matrix $\boldsymbol{P}$ of Eq. (4.45b) is singular.
(b) if the three points are aligned, then the same matrix $\boldsymbol{P}$ as in (a) is singular.
(c) State the necessary and sufficient conditions on the three points for the same matrix $\boldsymbol{P}$ to be invertible.

4.2 Show that a necessary and sufficient condition for the 4×4 homogeneous-transformation matrix $\boldsymbol{P}$ of Eq. (4.49) to be invertible is that the four points be non-coplanar—colinearity included in this case.

4.3 Using scientific software and as many affine transformations as you can, produce a plot showing instances of the conics described below:

(a) a circle of unit radius centered at the origin of the X–Y-plane;
(b) now, a hyperbola tangent to the circle, with the coordinate axes as asymptotes and focal axis at an angle of 135° with the positive direction of the X-axis;
(c) an ellipse tangent to the circle, with semiaxes of lengths 1 and 2, and focal axis passing through the origin and making an angle of 45° with the positive direction of the X-axis;
(d) a parabola, with axis of symmetry passing through the origin, at 45° with the positive direction of the X-axis, the parabola intersecting the Y-axis at point

Fig. 4.38 An ellipse lying on an oblique plane

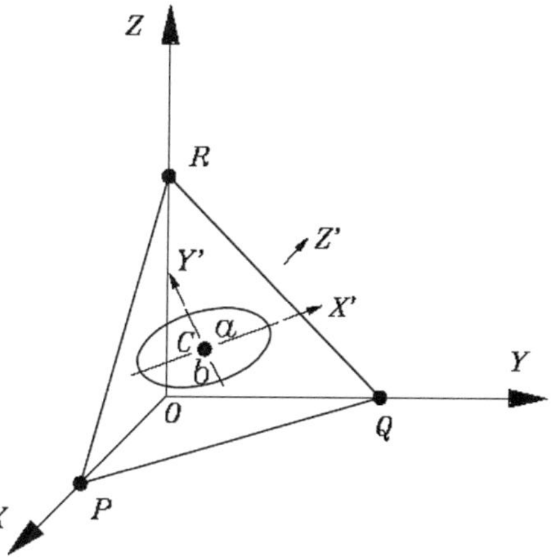

$I(0, 1)$. Moreover, the parabola intersects its axis of symmetry a distance $\sqrt{2}/2$ *below* the origin.

Include a brief description of the steps you went through in completing the drawing.

4.4 Rather than simply flattening the crests and roots of the screw of Fig. 4.30a, we would like to round them, but not using simply a circular shape. Instead, we would like to design rounds that blend with the sides of the triangle with G^2-continuity. Using scientific software, produce a plot of the thread profile, similar to that of Fig. 4.30b, with its rounded corners and G^2-continuity.
Hint: Think of Lamé curves, under a combination of suitable reorientation and non-uniform scaling. A Lamé curve of fourth degree would suffice. Think of distorting the circumscribing square, so that its sides intersect at the appropriate angle. Assume plausibly the location of the point of tangency of the distorted *Lamé curve with its circumscribing quadrilateral.*

4.5 Using a combination of: (i) *surface generation by revolution*; (ii) scaling; and (iii) Boolean operations, state the steps required to produce a conic journal bearing. This machine element has the shape of a frustum, similar to the one displayed in Fig. 3.8b, but with a central *through-hole* of axis coincident with the frustum axis. The dimensions of the frustum are diameter of the larger base, 60 mm; diameter of the smaller base, 30 mm; height, 20 mm; and diameter of the hole, 20 mm.

4.6 Obtain the affine transformation sequence that carries the unit circle centered at the origin of the X–Y-plane into the ellipse shown in Fig. 4.38. The lengths of the semiaxes of the ellipse are a and b, with $a > b$, the ellipse lying on the oblique face of a *rectangular tetrahedron* $OPQR$, with $\overline{OP} = \overline{OQ} = \overline{OR} = \ell$. Moreover, the center C of the ellipse coincides with the center of the oblique face, while the focal axis is parallel to the $\overline{PQ}$ edge.

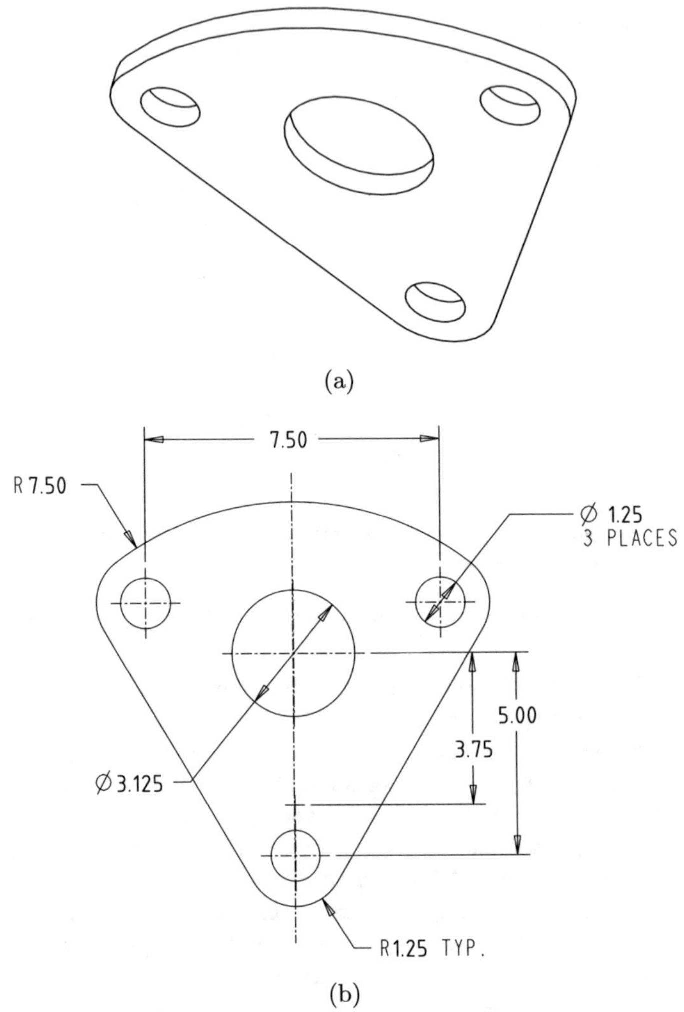

Fig. 4.39 A rod guide: **a** 3D view; **b** front view. All length in mm

4.7 Starting with the canonical equation of an ellipse, of semiaxes 10 and 15 units long,

(a) Plot the curve using the "implicitplot" Maple command, or equivalent in alternative *equivalent* software;
(b) Use a translation to shift the center of the ellipse to the point $P(3, 4)$;
(c) Next, use a rotation to turn the ellipse about its new center through an angle of 30°;
(d) Finally, use a reflection to produce the mirror image of the ellipse, in its new position and orientation, with respect to the Y-axis.

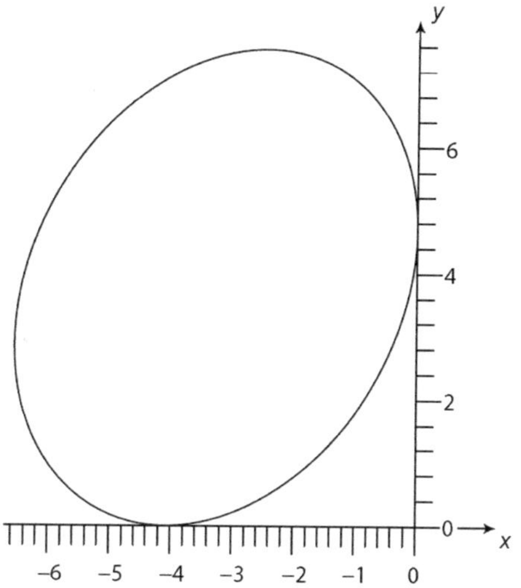

Fig. 4.40 An ellipsoid of revolution projected as an ellipse resting in a corner

4.8 Using CAD software, draw the rod guide of Fig. 4.39a with a thickness of 0.5 mm. To this end, use as many affine transformations as needed (e.g., down-scaling of the larger circle, then translating it, to simplify the work.) Then, use sweeping to produce the desired thickness.

4.9 Sweep the circle defined by

$$(x-2)^2 + y^2 = 1$$

around the Y-axis, to produce a torus. Plot the surface thus obtained.

4.10 *Producing a rhombus from a square*: Given a square with sides of coordinates $(1, -1)$, $(-1, 1)$, $(-1, -1)$, and $(-1, 1)$ produce a rhombus with equal side lengths, but with sides intersecting at 135° and 45°.
Hint: Think of affine transformations and, more specifically, of non-uniform scaling.

4.11 The design of an innovative machine calls for a special screw, produced by a generatrix having the shape of an equilateral triangle and a directrix normal to the plane of the generatrix.
Using a value of 25.0 mm for the triangle side and a pitch p of 10.0 mm, produce the screw using scientific software. Display a length of 50 mm of the screw thus designed and compare your result with what CAD software can give you.

4.12 Using scientific software, and nothing but affine transformations, produce the ellipse of Fig. 4.40, which shows the projection of an ellipsoid of revolution resting in a corner. The axes of the ellipse have lengths of 3 and 4 units. It is also known that the focal axis of the ellipse makes an angle of 60° with the X-axis. *Hint: Start with a circle of unit radius centered at the origin.*

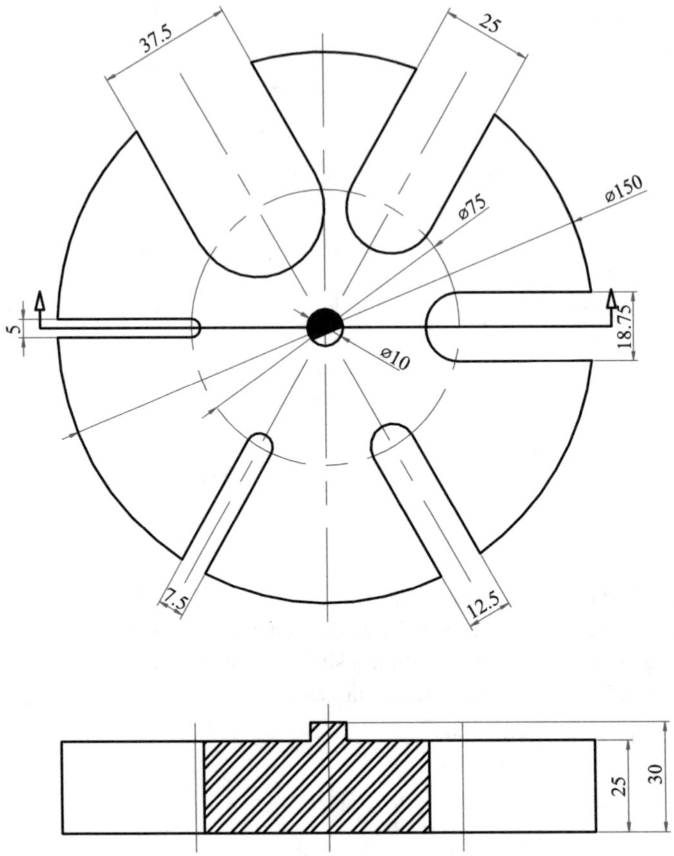

Fig. 4.41 The table of a hand press

4.13 Explain step by step how you would go about drawing the CAD model of the table of the hand press shown in Fig. 4.41. Suppose the CAD software tool you are using is very primitive and allows only for the set of commands given in Table 4.1.

4.14 The design of a machine structure requires an opening with the form of an equilateral triangle of side a, as shown in Fig. 4.42a. With the purpose of preventing stress concentrations, the triangle vertices have to be smoothed. A clever engineer has suggested to smooth the opening by means of a fourth-degree Lamé curve, as depicted in Fig. 4.42b, suitably distorted so that its tangents at the intersection with the X- and the Y-axes, intersecting with each other at point $C(1, 1)$, end up intersecting themselves at an angle of 60° at the same point C. To this end, you have been assigned the tasks below:

Table 4.1 Commands allowed by your CAD software tool

Generation	
`cylinder`(r, h, P)	Generates a vertical cylinder of radius r and height h with the center of its bottom surface located at P
`box`(w, d, h, P)	Generates a box of width w, depth d, and height h, with its vertex of smallest sum of coordinates located at P.
Transformation	
`copy`	Creates a copy of the selected object
`translate`(P, Q)	Translates the selected object using the same translation vector as that taking point P onto point Q
`rotate`(θ, P, Q)	Rotates the selected object by an angle θ around the axis going from point P to point Q
`pattern`(δ, n, P, Q)	Two possible situations: 1. If δ is a length, the CAD tool copies $n-1$ times the selected object, translating each new copy a distance δ from the preceding one, in the direction of the vector stemming from P and pointing toward Q. 2. If δ is an angle, copies $n-1$ times the selected object, rotating each new copy by an angle δ from the preceding one, taking the line containing points P and Q as the axis of rotation, with positive direction from P to Q.
`mirror`(P, Q, R)	Mirrors the selected object with respect to the plane containing the non-collinear points P, Q, and R
`scale`(σ, P)	Scales the selected object by a factor σ about point P
Boolean operation	
`union`	Merges all selected solids into one
`subtract`	Removes the volume common to the two selected groups of solids from the first selected group of solids
`intersect`	Keeps only the volume common to the two selected groups of solids

(a) Find the 3×3 homogeneous-transformation matrix $\boldsymbol{T}_d$ that will distort the Lamé curve as desired.

(b) Once the curve is distorted, it has to be scaled so that it will comply with the dimensions given in Fig. 4.42a, with a as a parameter, to be adjusted according to the various machine sizes planned. Find the 3×3 homogeneous-transformation matrix $\boldsymbol{T}_s$ that will do the required sizing.

(c) Once the curve is distorted and suitably dimensioned, it has to be displaced so that it will fit in the space allocated in the upper corner of the triangle of Fig. 4.42a. Find the 3×3 homogeneous-transformation matrix $\boldsymbol{T}_r$ that will do the required displacement. *Hint: Notice that, given the dimensions of the distorted and scaled Lamé curve, the required displacement reduces to one rotation about the origin.*

(d) Find the 3×3 total homogeneous-transformation matrix $\boldsymbol{T}_{tot}$ that will take the *quarter* of the Lamé curve lying in the *first quadrant* to its final shape

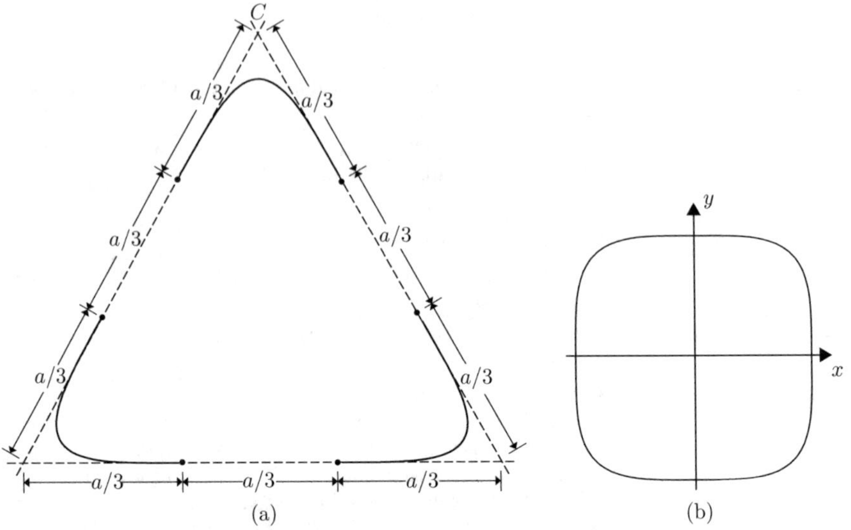

Fig. 4.42 An equilateral triangle with its corners smoothed—shown in **a**—by means of a fourth-degree Lamé curve—shown in **b**—suitably distorted, scaled, and displaced

and location in the upper corner of the triangle. *Hint: Let* $\boldsymbol{T}_1$ *and* $\boldsymbol{T}_2$ *be two homogeneous-transformation matrices. Their product is given below:*

$$\boldsymbol{T}_2\boldsymbol{T}_1 = \begin{bmatrix} \boldsymbol{M}_2 & \mathbf{t}_2 \\ \mathbf{0}^T & 1 \end{bmatrix} \begin{bmatrix} \boldsymbol{M}_1 & \mathbf{t}_1 \\ \mathbf{0}^T & 1 \end{bmatrix}$$

Hence,

$$\boldsymbol{T}_{tot} = \boldsymbol{T}_2\boldsymbol{T}_1 = \begin{bmatrix} \boldsymbol{M}_2\boldsymbol{M}_1 & \boldsymbol{M}_2\mathbf{t}_1 + \mathbf{t}_2 \\ \mathbf{0}^T & 1 \end{bmatrix}$$

4.15 An oblique duct is under design, to connect two identical circular bores of radius r and lying in parallel planes a distance h apart in the Z-direction, as shown in Fig. 4.43a, their centers separated by a distance $4r$ in the Y-direction. Find the 4×4 homogeneous-transformation matrix that distorts the right circular cylinder of radius r and height h, as shown in the same figure, into the duct shape. *Hint: choose your points before and after the shear judiciously, so as to ease the required matrix inversion.*

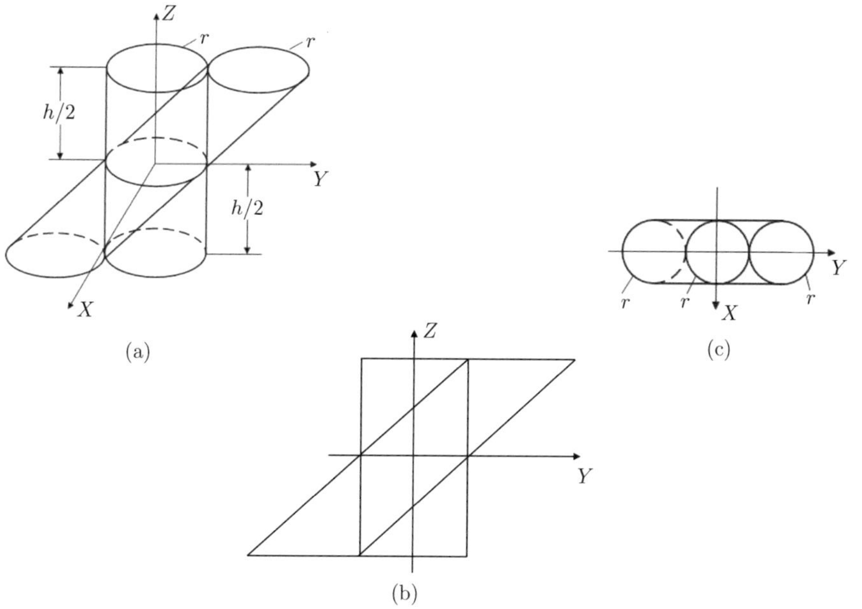

Fig. 4.43 A duct made of a distorted right circular cylinder: **a** isometric view; **b** top view; and **c** side view

References

Chaudhary M, Angeles J, Morozov A (2016) Design and kinematic analysis of a spherical cam-roller mechanism for an automotive differential. CSME Trans 40:243–252

Koenderink JJ (1990) Solid shape. The MIT Press, Cambridge

Singh S (2015) Design synthesis of custom-molded earphone sleeve. M.Eng. Project, Department of Mechanical Engineering, McGill University, Montreal

Strang G (1986) Introduction to applied mathematics. Wellesley-Cambridge Press, Cambridge

Index

J. Angeles and D. Pasini, *Fundamentals of Geometry Construction*, Springer Tracts
in Mechanical Engineering, https://doi.org/10.1007/978-3-030-43131-0

Printed in the United States
by Baker & Taylor Publisher Services